Moment of Truth

The Nature of Catastrophes and How to Prepare for Them

Kelly McKinney

A SAVIO REPUBLIC BOOK
An Imprint of Post Hill Press

ISBN: 978-1-68261-591-1
ISBN (eBook): 978-1-68261-592-8

Moment of Truth:
The Nature of Catastrophes and How to Prepare for Them
© 2018 by Kelly McKinney
All Rights Reserved

Cover Design by Christian Bentulan

posthillpress.com
New York • Nashville

Published in the United States of America

For Teresa C.

contents

foreword

In the spring of 1986, I was First Deputy Commissioner at the New York City Fire Department. Late one morning, word came down that the mayor was on hold for me.

"Joe, I need you at City Hall for a presser at three o'clock," he said.

For the past week, City Hall had been embroiled in a firestorm after a wave of corruption was uncovered in the City's Parking Violations Bureau, a wave that threatened to topple Mayor Ed Koch and his entire administration.

"I'll be there, Ed. What's up?" I asked.

"I'd like to announce you as the new Director of the Parking Violations Bureau."

I was stunned. This was arguably the toughest job in the world at that moment. Besides, other than how to feed the meters, I knew nothing about parking or parking violations.

"Can I have some time to think about it?" I asked. "It might be a good idea to run this by my wife."

"Joe, you're the only person I am asking to do this. It's OK if you say no. But you have to make the decision right now."

And so it was that I began a year-long stint within the swirling maelstrom that was the PVB scandal.

In my over three decades of public service, under three New York City mayors–John Lindsay, Abe Beame, and Ed Koch–I had seen my share of crises. And then, in 2004, City Hall called again. Michael Bloomberg asked me to serve as Commissioner of the newly chartered Office of Emergency Management. Over the course of the next ten years, I would add to a long track record as my agency responded to nearly five hundred emergencies every year, with large-scale incidents occurring on an almost monthly basis.

Unlike my Duty Team Chiefs (like Kelly McKinney), I was not on-call for just part of the time, I was "up" 24/7/365. Because the mayor owned every disaster, it was my job to own them for him. So I was standing behind him in a freezing Department of Sanitation garage on Bloomfield Street after the 2010 Christmas blizzard, and in dozens of other incidents like it, small and large, and very large. I stood behind him on 42nd Street after the Miracle on the Hudson, on Lexington Avenue after the steam pipe explosion, and on Yetman Avenue in Staten Island after it was devastated by Hurricane Sandy.

At OEM we were more than witnesses to history. At OEM we made history. It was a unique time for me, and for New York, as a never-ending stream of disasters taught us powerful lessons.

We learned that, in most cases, disaster avoidance is not going to be possible. The disaster you faced before will not be the disaster you face next, and that nobody, not even the mighty New York City government, can go it alone. It is important that we now share those lessons, not only in New York, but across the nation and around the world.

The lessons we learned, along with some great stories, are contained within the pages of this remarkable book. It is indeed a rare treat that one can read about a complicated and perplexing topic like when the "Big One" hits and come away with a straightforward how-to recipe. *Moment of Truth* offers the reader this gratifying opportunity and result.

In many ways this book is the legacy of my forty years of service to the people of New York City, including a decade as Commissioner of NYCOEM, the best municipal emergency management agency in the country, and perhaps the world.

—Joseph F. Bruno

prologue

an imagined catastrophe

polaris attack part 1 | *3:59 p.m.* | *Thursday, August 29* |
In the Atlantic off Brigantine, New Jersey

It's the last week of summer. Brigantine Beach is crowded with sunscreened bodies, colorful umbrellas, and toddlers digging in the hot sand.

Ninety nautical miles offshore, a two-hundred-foot-long cylinder of black steel rises slowly from the deep. A crew of thirty-seven sailors is aboard the *Gorea* ("whale"), a Sinpo-class ballistic missile submarine and the pride of the Pakorean People's Navy. Using a crude stealth technology, along with a bit of luck, the captain has managed to steer the *Gorea* to within striking distance of the American mainland.

Just below the surface, the captain orders the engines to a halt. He picks up the ship's phone and, after a brief exchange, places it back in its holder. He turns to his first mate.

"Launch when ready" is radioed back through the boat to the launch room in its stern.

All hands brace themselves and, after three long minutes, an explosion rocks the sub as its steam cannon blasts a sixty-ton projectile out of a hatch and upward toward the surface. The KN11/Pukguksong-1 (Polaris-1) ballistic missile jumps up out of the water; its rocket motor ignites, and it heads up into the sky.

The Polaris arcs high above Brigantine and its neighbors to the north: Beach Haven, Seaside Heights, and Belmar, its painted martial markings bright white against the matte black skin of the fuselage. After thirty seconds, the motor stops as the missile reaches the top of its arc. Its nose turns down and it begins its descent.

Silently it falls, northwest across Staten Island, the New York Harbor, and Lower Manhattan, toward streets jammed with delivery trucks, city buses, and taxicabs. A high school student from Ohio is walking with a tour group when he sees it dropping out of the clear blue sky. He points and, as he turns to tell a classmate, the missile finds its target. Ground zero is 41st Street and Sixth Avenue, one block from Times Square.

The payload aboard Polaris detonates on impact, and the world is changed forever.

impact | *4:04 p.m.* | *Thursday, August 29* | *RPG Studios* | *520 Eighth Avenue, New York City*

It's only a few minutes after four o'clock and you still have a lot of work to do. But instead of doing it, you decide to peek over the top of your cubicle, across that great expanse of office, toward the exit sign. The sight of some of your colleagues hurrying out flips the work switch in your mind to *Off*. Now the only thing you can think about is home.

But between you and home is that excruciating daily commute.

You feel a familiar anxiety as you ponder the race-walk through the streets, up Eighth Avenue to the Port Authority Bus Terminal. The 4:37 bus leaves well before the Mongol hordes of your fellow New Jersians. You won't catch it if you wait even two more minutes, so you grab your backpack and head for the door.

You're halfway there when the huge, dim room turns brilliantly white. A light, brighter than a thousand suns, bursts in through the windows, illuminating every tiny corner and crack; the air itself seems to ignite. Your eyes slam shut, but the light burns an image of veins through your eyelids.

At the same time, a long, low, boom explodes in your ears. The floor jumps and knocks you off your feet. As your body hits the floor, the windows explode.

Shards of glass fly across the great expanse of office floor, piercing your face and scalp and skin like tiny daggers. The ceiling crashes down…

ground zero | *4:04 p.m.* | *Thursday, August 29* |
41st Street and Sixth Avenue, New York City

The Polaris lands on the Fountain Terrace in Bryant Park just across Sixth Avenue from the Bank of America tower. At that instant, the device mounted in its nose unleashes a stream of neutrons at the uranium nucleus in its core, splitting the atom into fragments. This process, called fission, becomes self-sustaining as the neutrons released by the disintegrating atom strike nearby nuclei and produce more fission.

The chain reaction unleashes crushing waves of light, pressure, electricity, neutrons, gamma rays, alpha particles, and electrons in all directions: up into the sky and down the crowded city streets.

Radiation is first. Neutrons, gamma rays, and alpha particles, traveling at the speed of light, melt the aluminum off the sides of taxicabs, ignite tires on delivery trucks, and turn exposed skin to charcoal. Leaves on the trees in the park explode, bronze statues melt, and uncovered ground bursts forth into superheated dust.

Next is the light. For a split second, a fireball shines ten thousand times brighter than the desert sun at noon, searing the corneas of people as far away as Newark, New Jersey, and Greenwich, Connecticut. Traffic comes to a standstill as hundreds of thousands of cars, trucks, and buses across the city crash. Airplanes en route to Newark, LaGuardia, and Kennedy Airports fall from the sky.

At the Empire State Building a half mile away, light from the fireball melts the asphalt on 34th Street and burns the paint off its iconic television antenna. Inside the building, curtains and upholstery burst into flame. Its marble walls crack and pop.[1]

At the same time, a sharp, high-voltage spike of electricity sends a burst of electrons in all directions. This electromagnetic pulse melts the integrated circuits in every electronic device in Manhattan. The power grid and all communications systems are immediately destroyed.

The pressure comes as a blast wave and its leading edge, called the shock front, travels rapidly away from ground zero. It is a wall of highly compressed air traveling at 750 miles per hour and crushing everything—people, buses, buildings—in its path. Like all of the buildings around it, the nearby New York Public Library is ripped apart. Patience and Fortitude, the majestic stone lions that guard its entrance, are tossed into the air like leaves in a windstorm.

Last, but not least, is the heat. The heat from a chemical explosion (like that produced by dynamite or TNT), reaches several thousand degrees and creates a gaseous fireball. The plasma fireball created by a nuclear explosion reaches tens of millions of degrees. At this temperature virtually everything burns, first as individual fires and then, as massive winds drive them, they begin to merge into a single, gigantic fire—or firestorm.

Within seconds, Midtown Manhattan becomes a hurricane of fire. Heat from the firestorm forces gigantic masses of heated air to rise, creating a chimney effect that draws cooler air in toward the blast from all directions at speeds of hundreds of miles per hour. Anything that is not bolted down—hot dog carts, trash cans, pets—is sucked in.

The fireball creates a distinctive mushroom cloud that carries thousands of tons of highly radioactive debris particles several miles aloft. These particles are the "fallout." Heavier debris rains down nearby, but the lighter particles, like fine-grained sand, are carried by the winds and spread dozens, even hundreds, of miles downwind. Within minutes, everyone outside in the fallout areas receives a lethal dose of radiation.

The *Gorea* traveled underwater halfway across the world to bring forth hell on Earth. The skyscrapers of the great city center have been reduced to highly radioactive rubble. A million people are dead. Hundreds of thousands more will soon die.

the president is told | *4:24 p.m.* | *Thursday, August 29* |
Willard InterContinental Hotel | *Pennsylvania*
Avenue NW, Washington, D.C.

The aide-de-camp walks slowly across the stage to the podium as the president speaks to a packed auditorium. The commander in chief is not accustomed to being interrupted in midsentence and is infuriated as he is ushered backstage.

But at the same time, he senses that something is wrong. Just before he was interrupted, the crowd seemed to grow distracted and tense. He got a strange feeling—like static electricity—in the room. When people do not listen to him when he speaks, something is wrong.

As soon as he steps into a small conference room, another aide starts a slideshow of cellphone pictures. They show satellite imagery of a huge fireball and starburst-patterned blast wave radiating across the width of Midtown Manhattan. The president's annoyance grows.

"Can somebody please tell me what the hell I'm looking at?"

He is handed another cellphone, but there is silence at the other end.

"Hello? This is the president."

"Yes, of course. Sorry, Mr. President. Those pictures you see there were taken by our surveillance satellites within the last fifteen minutes. They are showing what appears to be a nuclear explosion in New York City. Based on those images, we estimate the yield of the device to be between five and seven kilotons."

There is a long pause. "Who did it?" the president asks.

"I'm afraid I don't have any information about that, sir. I would be happy to reach out to Langley for you if you like…"

The president is stunned. With the worst-case scenario clearly upon him, he is left speechless. Finally, he manages to blurt out, "So what are we doing about it? Like, right now?"

For a minute nobody speaks.

"We are trying to reach Secretary Ojami," somebody says, referring to the Homeland Security secretary.

"But right now we need to get you safe…"

first response | *4:25 p.m.* | *Thursday, August 29* | *FDNY 18 Squad Company* | *Greenwich Village, New York City*

The men and women of 18 Squad[2] are arriving back at the firehouse when they see the flash. Lieutenant Anthony Padilla and his firefighters jump off the rig and run to the front of the apparatus bay.

Their eyes scan east and west, but they see nothing but the deep blue sky. The lieutenant turns to see firefighter Tyesha Brown dashing to the corner for a view to the north. As her gaze turns up Sixth Avenue, she sees a massive column of fire rising a mile into the sky. She calls her mates over. They stand like statues and stare.

After pulling themselves together, they make their way back to the firehouse. The alarm bell is sounding, but all the radios are dead. 18 Squad doesn't need Dispatch to tell them where the job is. In three minutes they are back on the rig and headed north.

They make a grim passage up Avenue of the Americas, along sidewalks packed with dazed and bloody crowds of people. Most have no idea what has just happened or what they should be doing now. Storefront windows are shattered, and broken glass covers the street.

At 20th Street they pass the Limelight, a former Episcopal Church of the Holy Communion, now a nightclub. Memories of late nights there flash through Tyesha Brown's mind as she stares at the dark hole where the great rose window once was.

Stalled automobiles slow their passage, and the intensity of destruction increases as they near 23rd Street, about a mile from ground zero. Buildings are blown out, some have collapsed, and they can see the fires inside.

At this point, rubble and overturned cars across Sixth Avenue prevent any further progress on the rig. Padilla orders them to set out on foot, carrying their gear. Acrid smoke burns their eyes as they stumble out into the street.

As they walk, they are joined by other firefighters from Engine Company 33, Ladder Company 15, and Marine 1 along with police officers from NYPD Harbor Patrol. They look ahead toward Times Square and see a hundred-foot-tall wall of fire. The vision sends a shock wave of fear through them as they continue their grim trek north.

Things quickly go from bad to worse. The sidewalks here are filled with people, mostly sitting, some lying down. The radiation meters clipped to their belts show elevated levels, and the smell of gas from ruptured lines fills the dusty air. Smoke from the fires continues to thicken around them.

Collapsed buildings turn into buildings on fire, so they go to work, moving from burning building to burning building. Padilla decides that firefighting is pointless, so he orders his crew to drop hoses and search for victims. In the middle of the intersection of Sixth Avenue and 25th Street, a ruptured gas main burns with a roar, its fierce blue flame piercing the darkening sky.

part I | *collision course*

1 high intensity

we are increasingly vulnerable to catastrophe

"Welcome to the crossroads of the world."

The conductor's announcement comes loud and clear over decades-old speakers as the Number 1 train pulls into the Times Square subway station.

This is New York: international center of finance, technology, art, and fashion. It surrounds a vast natural harbor where the Hudson and East Rivers converge to meet the Atlantic Ocean. Sixty million people a year visit its landmarks, including the Empire State Building, Central Park, and the neon lights of Times Square.

It is the city that never sleeps. But for the eight and a half million people who live here, the vitality of the Big Apple has a downside, particularly when they try to get on with the mundane business of daily life.

the great seamless disaster part 1 | *Early afternoon* |
Saturday | Williamsburg | Brooklyn, New York City

As you tramp down the narrow staircase, you try to remember that moment when you first decided that you wanted to live in the big city.

You're venturing out from your tiny three-thousand-dollar-a-month apartment across town for a shopping trip that you know is going to take most of the day. The last time you braved the big-box store, you ended up with a lot of things you didn't need because you grabbed everything within arm's reach while you waited in a line that circled the store and backed up all the way to the front door.

As you emerge from your dingy tenement into the bright sunlight—sleep-deprived because your next-door neighbor watches action movies at all hours at full volume—you are hit with yet more noise: a wailing ambulance siren and a car horn honking in the street five feet away. You try to navigate the human flow of a crowded sidewalk. One would think that in a city of walkers, people would know how to walk. They do not.

You need some peace—some life essentials—like grass or fresh air or space. You're pretty sure they have those things in the suburbs, but not here.

invisible infrastructure

> *"There's not a single aspect of our lives—heating, food, fuel, communications, medical care, and the list goes on—that's not intimately tied to a network of machinery that can be readily disabled by a squirrel."*[3]

—Adam Stone, *Emergency Management* magazine

According to the United Nations, nine in ten city dwellers are exposed to "high mortality vulnerability" from disasters. And over the past fifty years, the population of cities, especially the megacity, has exploded.[4]

In 1950, there were two megacities with ten million or more inhabitants. In 2017, there were twenty-two, of which New York is but one. And New York is not just any megacity; it is the alpha megacity, with enormous assets and great strengths. The problem is that behind each of those great strengths lies great vulnerabilities.

New Yorkers live and work alongside an array of man-made, natural, and technological hazards, from terror attacks to climate change. Like all megacities, it is a *threat-rich environment.*

Eight and a half million people are packed into just three hundred square miles of dry land here...with a complex geography that rises a mere fifty feet above the waterline. Its three islands and peninsula are tied together by whisker-thin bridges and tunnels and an aging mass transit system that is breaking records for ridership.[5]

Like all megacities, it needs resources—electricity, food, fuel, water, data, transportation, money, healthcare—to keep it alive. A vast critical infrastructure has evolved to do this.

This critical infrastructure is an intricate web of interconnected systems—a *system of systems*—supercharged with massive computing power and complex mathematical algorithms. These algorithms create an artificial intelligence, or AI. Also known as machine learning, AI makes our critical infrastructure "smart." Smart systems use big data to analyze situations and to predict the future. They can sometimes even fix themselves. Yet beneath every bright control panel lurks a dark downside—because no matter how smart a system may be, it gets really dumb when it breaks.

the great seamless disaster part 2 | *Late evening* |
Saturday | *Williamsburg* | *Brooklyn, New York City*

You're back in your neighborhood and trudging up that cramped staircase. You decided to bite the bullet and pay the 150-dollar delivery fee, because there was no way you were going to get that new couch into a taxi and up those stairs by yourself.

You had a pretty good day. You chatted with a woman while waiting in the long checkout line (talking to a stranger...this never happens). She had just graduated from some college and was planning her escape from the big city. "Brooklyn is like a human Jenga pile," she said. "And how can anyone afford to live here? Yesterday I went to buy a tomato and was like, 'Wow.' "

Now it's late and you haven't eaten a thing since you said yes to the store-brand cinnamon roll. You decide to use Seamless to order Thai food from that tiny restaurant around the corner on Ten Eyck Street. (Seamless [also known as Grubhub] is a wildly popular virtual middleman that allows users to order restaurant delivery food without the major hassle of actually talking to a live person.) But for some reason the Seamless app isn't working. Twenty minutes go by as you try and try. And try again. Something is wrong, and you check Twitter:

"Seamless is down on a Saturday night?! #freakingout"

"Attention New Yorkers: Seamless is down. This is not a drill. Stay indoors, protect yourselves"

"It's Saturday night in New York City and Seamless is down. So...this is how it all ends"

All around New York City, human beings are gathering in small groups to stare into their cellphones. The only sound is the low grumble of stomachs. Eyes dart furtively around the room, as hungry, bad-tempered hipsters size up their companions for strength and will to live. Tonight, Seamless is down—and someone will have to die.[6]

You finally give up and forage in your tiny closet of a kitchen for something to fill the void in your stomach. Aside from last week's leftover Thai in the fridge, there is dry cereal (no milk) and half a bag of potato chips. You eat the chips, brush your teeth, and fall into bed.

high expectations

Seamless is an example of cutting-edge critical infrastructure that delivers to us astonishing levels of service and reliability.

If you managed to survive the Great Seamless Disaster, you could wake up in your cramped apartment, complain via Skype with a friend in Singapore, sell your Grubhub stock from a mobile app on the way to the airport, catch up on overnight Hollywood gossip on the plane, and get to San Francisco in time for a sun-drenched lunch in Union Square.

The reliability of these systems has set a high bar of expectation. The story of the Great Seamless Disaster is told with tongue in cheek since, aside from a brief disruption on a hot Saturday night in August 2015, Seamless servers never go down. So, we have come to expect extraordinary performance and reliability always. And for the most part we get them.

But there is a downside, because as a system evolves, its performance and reliability begin to diverge. To evolve it must become more complex, and that complexity breeds fragility. Increased performance becomes possible only with decreasing reliability. As expectations grow, so do the risks.

As a human being, my perception of risk is affected by many biases, including my emotions and my experience.[7] These biases pervade our society and sever our perception of risk from its reality. *This disconnect prevents us from doing what we need to do now to prepare for a dangerous future.*

only as strong as its weakest link

Seamless[8] is an online application that connects us to a spiderweb-thin *supply chain* that is embedded within a massive global logistics system. Supply chains stretch around the world and work around the clock to put shrimp into the refrigerators at restaurants and blood into the refrigerators at hospitals. Every one of these chains contains dozens of links, from farmers to factories to warehouses to retailers.

The global logistics system is the biggest and most complex beast imaginable. Held together by hundreds of thousands of interconnected databases, it is understood only by a small cadre of experts. And even that insight is flawed, with each expert's knowledge limited to their "link" and maybe a couple more on either side. Few comprehend the multitude of far-flung, fragile supply lines that span the globe, vulnerable to every kind and variety of risk: natural disaster, political unrest, or sabotage. Nobody knows how it all fits together or where all the landmines are. The only thing anybody knows for sure is that every one of these chains is only as strong as its weakest link.[9]

These are the circulatory systems through which flows the stuff that keeps society alive. The process is "lean," meaning it moves quickly. The shrimp from the farmer in Bangladesh gets to that Chinese restaurant around the corner in days, sometimes hours. Up and down the chain, links connect just-in-time. Nothing sits still for very long and nothing is stored, saving companies lots of money.

Innovators like Amazon.com and Alibaba continue to push the envelope of what's possible. You can get everything you will ever need delivered today, so you don't need to store anything either. This brand-new world is great for everybody. Until it isn't.

Like on that hot August night in 2015 when panic spread across the great city as the servers that powered Seamless' online applications suddenly went down. Hungry hipsters in the trendy neighborhoods of Manhattan and Brooklyn waited hours for orders that never came. As far as we know, most survived. But they were forced to venture out into the streets to hunt (i.e., talk to another live person) for food.

This will not always be so easy to do.

the most critical infrastructure

Power.

Every discussion of the risk to critical infrastructure must start here. Power is what we call a primary input; without it, nothing else can work.

It is supplied by a vast network of equipment that generates, transfers, and delivers the electricity that enables daily life. That network, known as the national grid, has grown increasingly complex and fragile.

It is aging; seventy percent[10] of the transmission lines and power transformers in the country are at least twenty-five years old.

It is massive, with 160,000 miles of high-voltage lines, five million miles of distribution lines, and tens of thousands of generators and transformers.[11]

*And it is tapped out w*ith virtually no spare capacity.

In New York City, the demand for electricity is so great that its limit is reached, and often exceeded, every summer. In July 2006, for instance, more than 175,000 people lost power in parts of Astoria, Sunnyside, and Woodside in Queens when the peak summer demand exceeded the ability of the network to handle it. I was managing the New York City Emergency Operations Center during that long, hot summer. Conditions got bad fast in the neighborhoods after the lights went out. We didn't know how long was too long without power in Queens, but based on that experience, I can tell you that it isn't much longer than a week.

In his book *Lights Out,*[12] Ted Koppel warns us that a major disruption to America's power grid is not only possible but likely. And when it happens, it will make 9/11 look like a picnic.

"Imagine a blackout lasting not days, but weeks or months. Tens of millions of people over several states are affected. There is no running water, no sewage, no refrigeration or light. Food and medical supplies are dwindling. Devices we rely on have gone dark. Banks no longer function, looting is widespread, and law and order is being tested like never before. Water towers on the roofs of high-rise buildings keep the flow going for two, perhaps three days. When this runs out, taps go dry; toilets no longer flush. Emergency supplies of bottled water are too scarce to use for anything but drinking, and there is nowhere to replenish the supply. Disposal of human waste becomes a critical issue within days."

—Ted Koppel, *Lights Out: A Cyberattack, a Nation Unprepared, Surviving the Aftermath risks to power*

Some say that the fragility of the electrical system is man-made, that we did it to ourselves when we deregulated what was a highly regulated industry. The explanation is simple and starts with the national grid itself.

Our great national grid has three jobs to do. First, it has to make electricity (we call that *generation*), then it has to move the electricity from where it is made to where people live (*transmission*), then, finally, it has to get the electricity into the outlets in people's homes (*delivery*). In the past, single companies were responsible for all of these jobs. But then deregulation and privatization overtook the industry and, over the past two decades, have combined to separate each of the three jobs from the others.

In the past, when your power was out, it was the utility company's job to get it back. Most people today can shop around for the best deal on electricity, but they are only buying it from someone who is buying it, as well. When the power goes out, the company you buy your electricity from will probably just shrug and point up the supply chain to the transmitter or the generator. So, as your power has grown cheaper, your risk of being left in the dark has skyrocketed.[13]

Another problem arises when the different computer and human systems that make up the national grid need to talk. Breakdown in communication was the cause of two of the world's worst-ever blackouts—the Northeast blackout in 2003 and the one across Italy and neighboring nations in the same year. Together, these power failures affected more than a hundred million people.[14]

our smart systems are making us dumb

The worry for disaster professionals is not so much the blackout as its aftermath, because there is little that government can do to improve conditions in hundreds of thousands of homes during a widespread blackout. During the Queens blackout in 2006, city officials handed out ice and bottled water and checked in on seniors and other vulnerable populations. And we watched and waited and worried until the lights came back on.

The problem is that most people have no idea what to do when the power goes out. Few of the people in those Astoria and Long Island City homes had a backup plan, or even understood the implications of seventy-two hours spent without the internet. Even fewer could hunt animals for their food. People with the ability to feed themselves without the Seamless app are becoming as rare as game birds in Williamsburg.

use it or lose it

These days, smart systems and "artificial agents" like robots perform a growing share of the tasks that human beings have always done; things like diagnosing medical conditions, driving cars, and writing technical reports. At the same time, we are gradually losing a wide range of these types of specialized skills. Osonde Osoba at the RAND Corporation calls this the deskilling effect.[15] This trend, a growing reliance on automation that saps our collective resilience, continues to accelerate. Massive investments by Google, Amazon, and others have resulted in breakthroughs in AI technologies that we have not even come to grips with. According to Osoba, the tendency toward automation in all aspects of 21st-century life will grow. The speed and scope of the socioeconomic impacts of AI will be "significant and unprecedented," he says.

As we spend our days satisfying the human cravings our modern critical infrastructure so marvelously and adeptly provides—like comfort and instant gratification and pleasurable distraction—we grow increasingly vulnerable to their disruption. At the same time, these systems grow increasingly vulnerable to disruption. The most susceptible, of course: the internet.

the cyberthreat

> **"Cyberthreats present a tremendous danger to our American way of life."**[16]
>
> —Former Secretary of Homeland Security John F. Kelly

There is one thing that all the systems that comprise our critical infrastructure—from power to hospitals to water to the supply chain to communications to retail stores to mass transit to banks—have in common. All are unalterably fused to that treacherous wilderness known as the World Wide Web.

The internet is a vast domain with many secret places that are not accessible using popular search engines like Google. These hidden places have names like Deep Web, Darknet, and Dark Web (we will refer to all of these as the Dark Web). You can only access the Dark Web with a special tool called a platform. Dark Web platforms give users access to sites that sell all kinds of bad things, like child pornography,[17] machine guns, and fake passports. One of the best known Dark Web sites was a marketplace called Silk Road that, at its peak, sold illegal drugs to more than a hundred thousand buyers.[18]

The Dark Web is also where cybercriminals go to get the tools they need to rob us. They use cryptocurrency like

Bitcoin to pay for malicious software ("malware") on Dark Web marketplaces, often in ready-to-use kits that promise quick payouts for those with little technical expertise. Target Corporation recently lost forty million credit card numbers to an attack that used an off-the-shelf malware kit purchased on the Dark Web for thirty-nine dollars.

One of the ways malware works is by breaking into your computer and taking it over, turning it into a robot (or "bot") that can be controlled remotely. Through command and control (or C2) servers, also purchased on the Dark Web, cybercriminals direct botnet attacks that spread malware to more devices, comb for credit card numbers, and mine for Bitcoin. These C2 servers do not rely on a single IP address and cannot be filtered or detected. They are also smart, meaning they can heal themselves if attacked and can reroute connections if one or more bots is taken down.[19]

a different breed of bad guy

Law enforcement experts admit their biggest worry is an attack to our critical infrastructure perpetrated via the internet. In this age of cyberwarfare, the only weapon the bad guys need is a laptop.

While the impacts of a cybercriminal attack are financial and can be devastating to their victims, for the megacity, the risk is systemic. The so-called threat trajectory for cybercriminality is extreme, with a growing gap between the power and sophistication of the attackers and our increasingly feeble ability to stop them. A different breed of bad guy—from hostile nations to "hacktivists" to terrorists—is using the protective cloak of the Dark Web to target the highly vulnerable control systems that run the national grid. In 2015, for instance, an attack on these controls—called Supervisory

Control and Data Acquisition, or SCADA, systems—took down parts of a power grid in Ukraine. A year later, Russian hackers targeted a substation, blacking out large areas of Kiev.

A large-scale attack on the national grid could send huge swaths of the country back to the Dark Ages. "It's not a question of if," says General Lloyd Austin, former commander of US forces in Iraq. "It's a question of *when*."[20]

Our brave new Internet of Things (IOT) world includes everything from cellphones to coffee makers, jet engines to oil rig drills. The problem is that most of these smart devices are equipped with little or no security. As more and more of our things are connected to the internet, more and more of us will be directly exposed to attack.[21]

And then there is the possibility of cyberattacks on the systems that control our nuclear arsenal. These so-called zero-day exploits could be used to launch missiles or disarm our missile defense systems. In a crisis, a president contemplating the use of nuclear weapons may not be sure that our command and control systems are secure. He or she could be forced into a decision to use nuclear weapons early or even to delegate the decision to military commanders in the field.[22]

the biggest threat to critical infrastructure

Of course, bad guys aren't the only threat to our critical infrastructure. After 9/11, disaster professionals in New York City focused on the risk of man-made threats, but time and time again, Mother Nature brought us back. Over the past two decades, New York City has endured an onslaught of natural disasters, including heat waves, blizzards, tornados, hurricanes, and even an earthquake.

Like most megacities, New York City is more likely than suburban or rural areas to experience natural disasters and is more vulnerable to their effects,[23] with floods, droughts, and hurricanes being the most devastating threats.

This is likely to get worse, with climate change exacerbating the impact of future catastrophes.[24] And—as we saw in Puerto Rico in 2017 after Hurricane Maria—natural disasters threaten not just individual systems but the entirety of a region's infrastructure.

what this book is about

> *"My job is to tell you things you don't want to hear, asking you to spend money you don't have for something you don't believe will ever happen."*[25]
>
> —Michael D. Selves, director of Homeland Security and Emergency Management, Johnson County, Kansas

In the disaster business, risk equals the probability of hazard times its consequences. The premise of this book is that in our ultra-modern society, all of the variables in this risk equation are higher than you think.

We are on a collision course with a range of natural and man-made threats. This is a hard reality, but these days, many people are getting a whiff of some sort of calamity ahead. Most cope by denying it. Some ease the anxiety with the belief that there are lots of people out there who get paid to make sure we are ready for whatever bad people or Mother Nature can throw at us. There is some truth to this. Disaster professionals across the country—in business, industry, and government—are working around the clock to prepare for disasters. However, there aren't nearly enough of us. And, as we shall see, some of us are working on the wrong things. The result is that as a nation, we are nowhere near ready for the worst-case scenario.

Everything in the following pages is intended to convey that insight: *You are not prepared for a catastrophe and assume that your government is—but your government isn't prepared for a catastrophe either.*

This book will explain why this is so and how we came to find ourselves in this predicament. It will start with the argument that everybody—not just you—is in denial about the threat of catastrophe. It will go on to tell the story of a worst-case scenario, dubbed "the mother of all disasters," to show you what it would look like if it happened today. It will then describe the "parallel universe" that the catastrophe creates and what it takes for disaster professionals to be effective there. Finally, it will give you some simple steps that you can take to prepare yourself and your family for catastrophe.

There is bad news here, about the state of our collective resilience and why our government isn't ready. But there is also a bit of good news. The good news is that some governments—we use New York City as an example—have the right system. A system that supercharges the government-led response to find and help the people who are trapped during a catastrophe, a system that is the model for the nation and for the world.

2 the brick wall of hope

we are oblivious to our vulnerability

home fire | *Middle of the night* | *Your bedroom*

You are having a nightmare. In your dream, you are lying in your bed surrounded by fire. You are paralyzed with fear as thick black smoke fills your nose and throat with the stench of burning wood and plastic. Somehow you realize that you are asleep and want only to wake up. And then you do. When you open your eyes, you see that the nightmare is real; your house and everything around you is on fire. The flames are so close that you are hot—hotter than you have ever been in your life. You feel the thick smoke against your skin as you struggle out of bed and fall to the floor. You remember that someone once told you to stay as close to the floor as possible. But there is no relief there. The heat and lack of oxygen cause your mind to race. Any hope you had starts to fade and you realize that this is the end. As you begin to lose consciousness, you hear—far off at the other end of the house— the high-pitched tone of a smoke alarm. A fresh battery for the nearby smoke alarm, the one you kept meaning to install but just never got around to—*the one that could have saved your life*—sits unused on your kitchen counter.

"let's roll the dice" is not a plan

It goes without saying that you should have put that new battery in that smoke alarm. But many seemingly obvious things are not so obvious before the crisis. I will start by stating the obvious now: *you need to prepare yourself and your family for disasters.*

It's not like you haven't heard this before. Ads urging you to build a kit or make a plan seem to be everywhere these days. From Ready.gov to the Centers for Disease Control and Prevention's (CDC's) Zombie Preparedness campaign to the American Red Cross's "Be Red Cross Ready" program, enormous time and effort are spent badgering you to be prepared. The problem is that much of this time and effort is wasted.

Why you should prepare is obvious. Let's start with the big one. While no one likes to talk about it, you could die. You should think about that for a second. Think about the impact that the death of someone in your immediate family would have on your life. Then think about the impact your own death would have on those around you. It's not just the emotional impact but also the practical one. Not having a will can cause significant emotional and financial stress for your survivors. If you haven't already done so, consider making a will and organizing your life insurance—they will help your family should the worst happen.

You could also lose your house and your valuables. You could be separated from family and close friends. You could lose your job or suffer big financial losses.

Believe it or not, disasters are not good for your health. Research shows that natural disasters worsen chronic disease, often because people lose access to their prescription drugs or treatment. Your mental health could be affected. You could lose your memory or your ability to concentrate. Worse yet, you could suffer post-traumatic stress disorder, depression, and anxiety, all of which would seriously impact your quality of life.

Finally, a disaster could cause major disruptions to your community—including loss of friends, neighbors, and social networks such as clubs or sports teams—leaving you isolated.

But you hope bad things won't happen to you, and you don't like to think about them. *This is why you don't.*

There are as many excuses for not preparing as there are good reasons to prepare, so why are you not doing it? I can think of a few reasons. Let's start with fatalism ("It is what it is. If my time is up, so be it"); fear ("Planning for a disaster will jinx me; if I do it, it will happen"); defiance ("I refuse to live in fear"); cost ("I can't afford it; I am not rich"); misplaced confidence ("The government will take care of me"[26]); complacency ("I am too old to start"); and faith ("God will protect me"). If I were to guess, I would say that the real reason is probably just good old-fashioned procrastination. You put it off, thinking you will get to it…eventually.

Here's the thing: you are not alone. Studies conducted over the past 15 years show that few of us are preparing.

Four years after Hurricane Katrina, only slightly more than half of those responding to a Federal Emergency Management Agency (FEMA) survey had any emergency supplies in their homes. Other national surveys report similar dismal findings. Although more than ninety percent of Americans think it's important, only about half admit to having taken any steps at all to prepare for an emergency.[27] Even people who live in higher risk locations, like earthquake or tsunami zones, don't do much to get ready.[28]

A uniquely human emotion called hope underlies our reasons for not preparing. Hope can be quite useful in our daily lives. For instance, what if instead of running off to work in the morning, you first took time to ponder all the bad things that could happen out there in the great big scary world? You might decide to stay home instead, curled up in a fetal position under a blanket.

While hope allows us to function in our daily lives unhindered by fear, it can be unhelpful too, preventing us from doing some easy things now that will improve our situation when something really bad happens.

You've heard the expression "Hope for the best and prepare for the worst." It's a good idea, but many of us only hope for the best and stop there. When it comes to the actual preparing part, we punt. It's too much work, too much stuff to buy, too many scary scenarios we don't want to contemplate. So, instead of contemplating, we block everything out with denial as impenetrable as a brick wall.

A brick wall of hope.

The wall gives us comfort. It lets us believe, "It probably won't happen to me."

because black swans

"Black swan" was a common term for an impossibility in 16th-century London, because nobody had ever seen a swan that wasn't white. This thinking changed in 1697, when Dutch explorers became the first Europeans to see black swans in Western Australia.

In his book *The Black Swan*, Nassim Nicholas Taleb blasts a hole in our brick wall of hope to illuminate this elusive creature. He presents World War I, the collapse of the Soviet Union, and the 9/11 attacks as examples. A former hedge fund manager and derivatives trader, Taleb argues that banks and trading firms are vulnerable to black swan events and that they are exposed to losses far beyond anything predicted by their defective financial models.

When Taleb's book was published in 2007, disaster professionals in New York City, and around the country, were struck by its unique insights about risks on Wall Street and their analogies to the wider world of New York City and beyond.

Since then, disaster professionals have adopted the term "black swan" to refer to widespread, catastrophic disasters. That is how we will use the term in this book: as synonymous with "catastrophe."

Taleb's black swans have three characteristics. First, they are way beyond our expectations, because nothing in the past points to them being likely to happen. Second, they have massive impacts. Third, we create explanations for them after the fact, making them seem predictable.

The probability that a black swan will occur where you are today is low—so low that it cannot even be calculated. Therefore, you ignore it. Yet it seems to happen all the time.[29] And when it does happen, its consequences are enormous—so much so that it plays a disproportionate role in human history. Taleb's point is not that we need to be better at predicting black swans. What he wants us to understand is that they are much more frequent, and destructive, than we think.

the end of the world as we know it

Not everyone needs to be convinced of this. In *Lights Out*, Ted Koppel tells the story of the "preppers." These are the people who live beyond the brick wall of hope, actively preparing for nukes and massive blackouts by stockpiling food and ammunition.

Preppers know in their hearts that the catastrophe is right around the corner. They create special terms to describe this and to create a group identity. These include acronyms like YOYO (you're on your own), GOOD (get out of Dodge), and, of course, TEOTWAWKI (the end of the world as we know it).

Among the millions of preppers across the nation, Koppel introduces us to a Wyoming homesteader who made thousands of adobe bricks for his house by hand and dug a three-acre lake that he stocked with fish.

The Mormons were preppers before prepping was cool. Koppel shows us the unrivaled disaster preparedness of the Mormon Church, with its enormous storehouses, high-tech dairies, orchards, and private trucking company—the fruits of a long tradition of anticipating the worst.

who owns the black swan?

Most of us live our lives surrounded by a brick wall of hope, while others prepare for an imminent end. The reality is that people are either inclined to prepare themselves and their families or they are not. That is just human nature.

But it cannot be true of the people, like me, who get paid to do these things.

You depend on us to be ready to go to battle with the black swan. In the disaster business, we call that "owning the problem." But because we are people too, disaster professionals inevitably struggle with that same brick wall of hope. So, as you point at us, we spend precious time and money creating preparedness campaigns that point right back at you.

In New York City, our brick wall of hope fell sixteen years ago, at 9:59 on a beautiful September morning, when the South Tower of the World Trade Center collapsed and ripped a hole in the universe.

I watched it from a third-floor window at the Department of Health headquarters in Lower Manhattan. As the floor shook and the room reverberated with a thunderous staccato roar, I felt the social fabric giving way. It seemed to me that we had reached that imminent end.

From that morning on, New York disaster professionals have had no illusions about our vulnerability to catastrophe. We gained a gift on that terrible day, a gift of insight. Far from making us fearful, it was empowering, and we have worked hard to keep a tight hold on it. Everything we have done since has been based on the knowledge that we are exposed to impacts far beyond anything that we can ever be ready for. We can never prepare enough or practice enough or know enough. We can never be strong enough to battle the black swan, so we can never stop building—our team, our plan, our resources, and our capability.

After I joined the New York City Office of Emergency Management (hereafter referred to as *OEM*) in 2006, we worked to bring that gift of insight to our colleagues around the region and the nation. We spread the gospel of urgency to surrounding counties and states and to the federal government—not only to help them, but to make them better able to help us.

As we shall see, we made real progress. Although it was a constant struggle. There were many challenges—complex problems to solve, resource constraints, and politics—but most of the challenges were between our ears. Or, more specifically, between the ears of the people from whom we needed help.

our imaginary friend

We faced the same issues time and again: passive resistance, confusion, and ignorance. It got to the point that we knew exactly what people would say and when they would say it. As strange as it sounds, we made up an imaginary friend to explain this phenomenon. Our imaginary friend's name is Bruce.

Bruce is a buff (a "buff" is a slang term for an enthusiast—one who likes the world of emergency response and feels comfortable amid disaster). He is just under six feet tall and sports a noteworthy paunch beneath a polo shirt embroidered with a stylish logo. His wide, tanned face is dominated by a furrowed brow and an enormous gray mustache. Around his neck is a lanyard clipped to a handful of shiny cards.

He has a first responder background and lots of experience. This means he is the person you want beside you in the field when all hell breaks loose. Nothing stops him from saving lives—not iron doors or bureaucratic red tape—when the chips are down.

He is a sociable fellow. He likes to laugh and joke but, at the same time, has strong opinions about things. For instance:

Bruce is confident. He has seen a lot of "jobs" (i.e., disasters) and has developed a keen instinct. He relies on that instinct to make decisions, and that has worked well for him. He knows what will happen and what he will do when it does.

The good news for us is that right now, his gut is telling him that we are in good shape. He doesn't see any warning signs, no imminent or inevitable disasters, on the horizon.

Bruce has perfect 20/20 hindsight. At some point in the aftermath of every disaster, he convinces himself that he saw it coming. Most of what went wrong that led to the disaster is perfectly explainable.

Bruce loves predictability and hates randomness. Like most people, he interprets random noise as valuable data. He refines all new information to create his predictions of the future. Even though disasters are random and unplanned, he sees them fitting into a clear historical progression.

Bruce is slow to recognize the new. Even though he lived through 9/11 with us, he does not believe in the black swan. He still thinks, as Taleb says, "within the parameters of the bell curve" and ignores large deviations. So, when the black swan comes, he will be surprised by it and shocked by its extreme impacts. But that's okay, because ultimately it will be made predictable again by the explanations he concocts afterward.

Bruce hates complexity. He is a simple man who knows what he knows and that's that. He doesn't much like new ideas. Ideas remind him of effeminate philosopher types who sit around in salons smoking French cigarettes, never actually doing anything. Besides, ideas are too abstract and require too much thinking.

Bruce likes stories. He runs away from ideas and embraces stories instead, especially war stories. A war story is a memorable personal experience, typically in a disaster. Bruce uses war stories[30] as blocking mechanisms against complexity and will trot one out at the first sign of a complex discussion (as soon as the thinking wheels in his head start to turn). Sometimes they are firsthand experiences, but often his war stories are merely hearsay.

Of all of Bruce's biases that OEM battled in post-9/11 New York City, the most fearsome was the story. Stories are good for learning, because they can communicate ideas in a user-friendly way. Stories can be dangerous, too. They breed complacency. They make us think we understand something that we don't understand at all.

A bad story is like a rock in the head, hard and immovable, taking up space needed for thinking. The mission of this book is to describe the bad stories and call them out as myths. At the same time, it is chock-full of good stories to replace them.

We will start with a bad story, one of the most egregious myths in the disaster business: the myth of the Whole Community.

the myth of whole community

In this popular story, preparing for disasters such as natural disasters, acts of terrorism, and pandemics is not the responsibility of government. According to the Whole Community myth, everyone[31]—individuals and families, businesses, faith-based and community organizations, nonprofit groups, schools, you name it—is working together, hand in hand, to build a resilient nation.

If this sounds too good to be true, that is because it is.

These days, individuals and families, businesses, faith-based and community organizations, nonprofit groups, and schools have a lot on their plates. Full plates—with things like fatalism, defiance, cost, misplaced confidence, complacency, faith, and good old-fashioned procrastination—thwart real progress in preparedness.

And, like all myths, the Whole Community myth contains a grain of truth, because there are plenty of people working to be ready for disasters. It's just that the idea that it is happening spontaneously everywhere, in an organized way, to increase our collective resilience is a fiction. It is classic muddled thinking to say that everybody is doing something, since it is the same thing as saying that nobody is.

Whole Community is a story made up by disaster professionals. Some say it exists so that we can avoid responsibility; instead of pointing to ourselves as responsible to lead preparedness for the nation, it's a whole lot easier just to point back at you.

the story of the turkey

Good stories are invaluable, because only a good story can displace a bad one. An example of a good story is Taleb's Story of the Turkey:

> "Consider a turkey that is fed every day. Every single feeding will firm up the bird's belief that it is the general rule of life to be fed every day by friendly members of the human race "looking out for its best interests." On the afternoon of the Wednesday before Thanksgiving, something unexpected will happen to the turkey. It will incur a revision of belief."

The Story of the Turkey is not good in the sense that it contains good news. It is a good story because it teaches us a critically important lesson: *what has happened in the past has no bearing on what will happen in the future.* The turkey's confidence increased as the number of friendly feedings grew, and it felt increasingly safe even as its imminent slaughter drew ever closer.

If the turkey survives until the next day, it means that either a) it is immortal, or b) it is one day closer to death. If we survive until tomorrow, it could mean that either a) our critical infrastructure systems are infallible and our public safety systems are impenetrable, or b) we are one step closer to catastrophe.

Both conclusions rely on the same data.

we are all bruce

Bruce has a lot of things going for him. He is good with technology tools like drones and radios. And nobody can drive an emergency vehicle better. But his biases stand in the way of our readiness. In every encounter, every conversation, we must work twice as hard just to overcome them.

By now you may be asking yourself, "Who is this Bruce person?"

For OEM, Bruce was the executive director who wouldn't commit his agency's people or resources to the cause; the borough commissioner who complained that disaster exercises were a waste of time; and the fire chief who "knew" that the worst-case scenario would never happen.

The truth is that Bruce's biases are part of standard-issue human frailty, so there is a little bit of Bruce in all of us. Bruce is the soccer mom who drove into a flooded intersection (with a bumper sticker on her minivan that reads, "Turn Around Don't Drown"). He is the Williamsburg hipster who didn't bother to replace that smoke alarm battery, and he's the senior citizen in South Florida who didn't leave his beachfront condo as the hurricane approached.

Talking about Bruce's faults and recognizing them in ourselves are the first steps in the never-ending battle against complacency. The critical next step is to create a mindset that counters those biases.

Disaster professionals must recognize, for instance, that complex discussions are essential, because the black swan is a complex beast. We must embrace ideas as well as stories because problem-solving is impossible without ideas, and we cannot hope to confront the tsunami of problems that the black swan will bring without them. We must replace our misplaced confidence with a "we don't know"[32] attitude, because the only thing we know for sure is that there is so much that we don't know—like when the black swan will come or what its impacts will be. That is why we can never prepare enough and can never stop building our resilience and our capabilities.

We must be open to the unexpected, instead of being shocked every time it happens. Novelty is a particular skill of the black swan, and we must embrace it. We must break down that brick wall of hope in our minds and *know* that the disaster is coming, so that instead of debating probabilities, we can concentrate on consequences.[33]

Finally, we cannot take an absence of proof to be proof of absence. Just because the ship's lookout in the crow's nest is not crying out, "Thar she blows" does not mean that our black swan is not just coming into view over the horizon.

shit happens

After 9/11, it was obvious to many in the intelligence community that an attack had been imminent. After Hurricane Katrina, it was obvious that the infrastructure of New Orleans had been long overdue for strengthening.[34]

According to Taleb, when we explain black swans as the product of some motivating incident or series of incidents, we are being fooled by randomness.[35] It is foolish to think we can understand the causes of catastrophe, even after the fact. After all, scientists have failed to accurately predict the timing of earthquakes and terrorist attacks even though considerable effort has been spent trying. Discovering the cause of a catastrophe is mostly an exercise in hindsight.

Despite this, scientists and media pundits continue to think that everything that happens in the world has a cause. They believe that every cause has an effect, and effects can be traced back to their causes.

You probably think the same way. When bad things happen, you ask why. You look for a cause, the explanation for why things went wrong. And you think that if you can understand that cause, you can prevent the bad thing from happening again.

Disaster professionals in New York City know this to be a myth. We call it the "myth of the cause." After fifteen years and hundreds of disasters—small, large, and extra-large—we, like Edward Aloysius Murphy, Jr. (of Murphy's Law[36] fame), know the truth. And the truth is that *anything that can go wrong, will go wrong.*

We call this truth "*shit happens.*"

Interestingly enough, there is a scientific basis for the idea that shit does, in fact, happen. It is a theory that maintains that for complex systems, like cities or critical infrastructure or even our own atmosphere, shit just happens. This theory proposed by Danish physicist Per Bak[37], is based on a property in physics he calls "self-organized criticality."

The theory of self-organized criticality rejects the notion of cause and effect; essentially, it says that an inevitable catastrophe is embedded within every complex system. Political systems, by their very nature, lead to terrorist attacks; financial systems lead to stock market crashes; electrical power grids experience thousand-year blackout events once a decade; and cities experience five-hundred-year flooding events once a year.[38] Take, for instance, the complex systems within the nuclear power plant at Three Mile Island.

the normal accident at three mile island

In March 1979, a mechanical failure[39] at the Three Mile Island Nuclear Generating Station in Middletown, Pennsylvania, resulted in a loss of coolant in its Number 2 Reactor (or TMI-2). This led to a partial meltdown of the core and a release of radioactive gases into the atmosphere. The incident was so severe that it was rated a five on the seven-point International Nuclear Event Scale.

When it was over, several state and federal agencies mounted investigations into the crisis, the most prominent of which was the President's Commission on the Accident at Three Mile Island. Most concluded that this disaster, like so many others, was caused by many small failures that cascaded into a catastrophic failure.

Babcock & Wilcox, the company that built the reactor, called the accident "discrete." They said that it started with an equipment failure and was compounded by a failure by the operators to follow the correct procedures. If the operators had done their jobs, they said, the accident could have been avoided.

By labeling the accident "discrete," Babcock & Wilcox was displaying classic black swan behavior: it was creating an explanation for the accident after the fact, making it seem predictable. Someone made a mistake, they argued: the equipment failed, the design was flawed, but something can be done about that. The problem can be rectified.

Others see this as delusion. They argue that the equipment failure was minor, and the operators did just what they were supposed to do. Charles Perrow, professor of sociology at Yale University, describes the event as "mysterious" and "incomprehensible" even to the Babcock & Wilcox experts who were at the site at the time of the accident.

Perrow argues that the disaster was caused exclusively by the *complexity* of the highly connected systems at the plant;[40] that it was *normal*, even *expected*.

It was bound to occur and is bound to occur again. This is because it emerged from the characteristics of the system itself. It cannot be anticipated, and it cannot be prevented. Furthermore, it is impossible to design or build in such a way as to anticipate all eventualities in complex systems where the parts are tightly coupled.

Worse yet, these types of "normal accidents" are incomprehensible when they occur. That is why operators usually assume something else is happening, something that they understand, and so they act accordingly. By being "incomprehensible," normal accidents are uncontrollable.

Safety systems, backup systems, quality equipment, and good training all help prevent accidents, but the complexity of systems outruns all controls.

myth of the cause

The myth of the cause teaches us that because we cannot see a cause does not mean that we do not have to worry about the black swan. The highly interconnected systems that comprise our critical infrastructure are far more complex than those at TMI-2 forty years ago. Every complex system contains the seeds of its own destruction; each is moving toward the precipice of catastrophe rather than away from it, by its very nature. Not even a tiny, random tremor is needed to trigger a major collapse—unexpectedly and resoundingly.

And when accidents occur in high-risk systems, such as those dealing with toxic chemicals, artificial intelligence, or nuclear weapons, the consequences can be catastrophic.

According to Jeffrey Lewis, a scholar at the Middlebury Institute of International Studies at Monterey, our nuclear weapons stockpile is "an increasingly elaborate machinery for destruction that [is] growing too complex for existing human systems to control or even for a single human being to fully comprehend."

> *"Technology is certain to disrupt nuclear deterrence. The only question is whether this means the end of nuclear weapons or the end of us."*
>
> —Jeffrey Lewis[41], "Our Nuclear Future"

welcome to extremistan

In *The Black Swan*, Taleb describes the imaginary worlds of Mediocristan and Extremistan.[42]

Mediocristan is the world normal people live in and where normal things happen. In Mediocristan, things are predictable and expected. Low-impact disasters are more likely to occur, while huge disasters with massive impacts are highly unlikely.

Extremistan, on the other hand, is the world where unfairness reigns and the unexpected happens. In Extremistan, nothing can be predicted accurately, and events that seemed unlikely or impossible occur frequently and have a huge impact.

In New York, our Mediocristan approach to disasters evaporated in the Extremistan reality of the post-9/11 world.

Now, in the wake of divisive elections and polarized populations, when the threat of armed conflict and nuclear war are at an all-time high, fate is poised with her hand raised ready to strike us a shocking blow.

The risk is here. Like static electricity, you can feel it in the air. Mediocristan is history. Ladies and gentlemen, this is Extremistan.

what all of this means for you

By now you may be asking yourself, "What does all of this mean for me?" It means a lot, actually. And it starts with the idea that you should begin to peer out from behind that brick wall of yours and think for a minute about the many different varieties of disaster, small and large—and very large—that could affect you. Or that could affect your family, your neighbors, your city (or megacity), your state, and your nation.

Because someday you will find yourself amid one of those disasters. That will be your moment of truth: a painful time that will come with a gift of insight—insight about the mistakes you made and the actions you did not take that would have increased your options or maybe even saved your life.

Although you are responsible for these actions for yourself and your family, others are responsible to act, too. You should worry a little bit about them because, even though disaster professionals project a lot of confidence, they are worried, and more than a little bit.

They are worried about their moment of truth: that painful moment when they are first engulfed by the black swan. When they realize that they, like everybody else, are overwhelmed by it, that there is too much to do and too little time to do it, that there are people waiting for help that is days away, especially the vulnerable, elderly people trapped in their homes without heat and disabled people with no electricity to power the devices that they need to survive.

"Wait a second," you might be thinking. "I have my own problems. Why should I worry about theirs?"

Because when the black swan comes, it could be you and your family or your friends or your neighbors who need help—the help that can come only if all disaster professionals across the nation come together. And that kind of help can only happen if all disaster professionals across the nation come together now to prepare to do it.

In this, the richest and most powerful nation the world has ever seen, the problem is not resources. We have huge resources, incredible technology tools, and plenty of smart people. There is only one problem and that is complacency. This complacency, which prevents us from preparing for the next black swan, is everybody's problem.

3 moment of truth

the mother of all disasters

What if we could get a glimpse into the future?

What if we could see all of the disasters that could happen and all of the bad things that they could bring?

We could then add those bad things together—things like massive flooding from hurricanes, collapsed buildings from earthquakes, empty store shelves and freezing homes from power failures, and deaths from bioterror attacks.

What would that look like?

To be sure, the combined impacts from those disasters would be enormous, even catastrophic. But the impacts from all those disasters combined pale in comparison to the devastation produced by even one nuclear explosion.

The consequences of a nuclear attack are so widespread, so pervasive, and so enduring that it must stand alone as our single greatest threat.

It is the mother of all disasters...

an imagined catastrophe (continued from the prologue)

escape from new york | *4:49 p.m.* | *Thursday, August 29* |
RPG Studios | *520 Eighth Avenue, New York City*

You wake up to yelling as your coworkers find you and help you out from under the collapsed ceiling.

Nobody knows what happened, but somebody has looked outside and says that it is "definitely not safe." Eleven people are trapped with you in the office, thankfully nobody is seriously hurt.

Cellphones and landlines are dead, but somebody finds a battery-powered transistor radio. Static fills the room as your coworkers watch you move the needle back and forth across the dial. You manage to find a station at 90.7 on the FM dial. WFUV is broadcasting from the Fordham University studios on its Bronx campus via an antenna atop nearby Montefiore Medical Center. Clearly shaken, the announcer intones everything she knows about what is going on. Hope fades as the enormity of your situation begins to sink in...

"A nuclear explosion has occurred in Manhattan. Thousands of people are feared dead. Many thousands more are injured, including many first responders. Thousands are evacuating the city by walking across bridges and tunnels into New Jersey. Power, television, and telephones are out. Streets and highways in and around New York City are blocked with abandoned vehicles. The Brooklyn Bridge, Brooklyn Battery Tunnel, Holland Tunnel, Lincoln Tunnel, and Manhattan and Williamsburg Bridges are closed. Mass transit, including all trains and buses, is closed until further notice. Nearly every hospital in Manhattan is closed. NYU Langone and Bellevue Hospital are open but overrun with the injured. Governors in New Jersey and Connecticut have declared states of emergency. The governor of New York and the mayor of New York City are missing and presumed lost. The New Jersey health commissioner recommends that everyone in and around New York City shelter in place until further notice."

You and your colleagues discuss your situation. Not knowing what else to do, you decide to shelter in place and wait to be rescued. You work together to seal the windows and forage for food and water. The accountants set up a system whereby everybody shares the meager rations fairly.

Within a day, the radio batteries die, and you begin to feel cut off. Your colleagues grow restless. Some decide to leave the sanctuary of the office. They begin to venture out alone in groups of two or three. Nobody comes back.

Two days pass and still help doesn't come. Based on the reports you heard on the radio, you're pretty sure it never will. The only thing you can think about is getting home. So, you decide to strike out on your own.

Outside, the sidewalk where you have walked every day for nearly eight years is utterly foreign. Your senses are assaulted by the stench and the eerie silence. Eighth Avenue, once loud and lively at all hours of the day and night, is deserted. The storefronts are dark and empty.

You know that the George Washington Bridge, nearly eight miles north, is the only way across the Hudson River, so you begin the long walk. The deserted sidewalks are nearly impassable with litter and debris. Everywhere you look, you see bodies. There are plenty of cars on the streets, but none of them are moving; most have crashed into storefronts.

As you skirt the fringes of the blast zone, you stare at the utter devastation within it. Fires are burning, smoke fills the air. You try to recognize any landmarks and finally you spy the remains of the UN's headquarters on the edge of the East River. You realize you're looking all the way across the island of Manhattan—everything from the Hudson River to the East River is gone.

You see a bus at a bus stop. It is full of people. Some are standing; some sitting. All are dead.

You walk along the river and past the piers. Normally, you would see the Hudson River between the piers. Now you can't see any water there; instead, all you see are dead people. There is no space for the water, just dead people floating there.

confusion at the top | *12:49 a.m.* | *Monday, September 2* | *Secure Situation Room* | *White House, Washington, D.C.*

The president is talking with a group of his secretaries. Three days have passed and still he can't get a clear picture of what is happening.

The stories coming out of Manhattan are the stuff of nightmares. Most of Midtown continues to burn. Hundreds of thousands, perhaps millions, are feared dead. There is an urgent need to organize and start dozens of different rescue operations, including search and rescue and medical evacuations. But very little is happening. Nobody is quite sure who is responsible to lead them, and the scene is deemed too unstable and dangerous to guarantee the safety of the teams. Most are confining their efforts to the periphery, for fear of putting rescuers in harm's way.

The Army and National Guard are dropping supplies from the air into makeshift triage centers, but it seems too little and too late.

New Jersey and Connecticut have been overrun with people, but Delaware, Pennsylvania, Rhode Island, and Massachusetts have closed their borders, sealing the Tri-State area off and leaving it to its fate.

FEMA has established a massive joint field office in an empty shopping center in suburban Maryland. It is a frenzy of activity, with representatives from federal agencies, non-profits, the private sector, and academia. But as far as the president can tell, it has no connection with anybody on the ground at the disaster site. He is getting its situation reports, but they tell him less than social media and television news.

They are showing gruesome scenes of the dead and dying in the fallout zone; and of the thousands scattered in nearby towns and suburbs, sleeping under tarps and on the open ground.

Desperation is growing and his military generals are clearly frustrated by a lack of clear orders. They are reporting dissension in the ranks, with officers threatening to strike out on their own.

So even though uncertainty is an unfamiliar feeling for him, for three long days the president has been confused. And then, ever so slowly, as the long night ends and the dawn begins, clarity takes hold in the form of two truths. The first is that the situation is bad—far worse than his advisors had been telling him. The second is that his government is completely overwhelmed, with no ability to control events.

no answers, only questions | *Noon* | *Monday, September 2* | *East Room* | *White House, Washington, D.C.*

The muffled quiet of the packed room is broken by the whirr and click of the cameras as the president steps up to the podium at the start of the press conference.

"I just received an update from the Homeland Security secretary, Secretary Ojami, and the other Cabinet secretaries involved, on the latest developments in New York City. I can tell you that our thoughts and prayers go out to the people of

New York City and others affected by this enormous tragedy. There is so much that we don't yet know, but here is what we do know now.

"Last Thursday, at approximately four o'clock in the afternoon, a missile was launched from a submarine that had invaded our territorial waters off the Eastern Seaboard of the United States.

"The missile was carrying a nuclear device with an explosive yield of more than 10,000 tons of dynamite. As we all know, that missile landed near New York City's Times Square. The device onboard detonated immediately, and most of Midtown Manhattan was destroyed.

"We don't yet know the perpetrators of this unspeakable evil, but its effects have been devastating. We do know that thousands, possibly hundreds of thousands, have been killed and many more have been injured and urgently need medical care. Response efforts are hampered, however. The city is cut off, with all roads, bridges, and tunnels closed.

"We are dealing with the worst disaster in our nation's history, and that's why I have called the Cabinet together. The people in New York City and the rest of the United States expect the federal government to work with the state government and the local government to launch an effective response. That is why we have been attempting to contact those elected officials both in the city and in New York state.

"The United States Coast Guard is conducting search-and-rescue missions off the New York coast. The Department of Defense is preparing to deploy major assets to the region. These include the USS *Bataan*, to conduct search-and-rescue missions; eight urban search-and-rescue teams; the Iwo Jima Amphibious Readiness Group; and the hospital ship USNS *Comfort*, to help provide medical care. FEMA and the Army Corps of

Engineers are working around the clock to try and assess the damage and determine a course of action and next steps."

The noise and chatter in the room grow. The president is talking about the crisis, but his words do not resonate. The nation has questions that need answers. The president presses on...

"Our second priority is to sustain lives by ensuring adequate food, water, shelter, and medical supplies for survivors and dedicated citizens—sorry, for dislocated citizens. FEMA is moving supplies and equipment near to the hardest-hit areas."

At this point an exasperated television correspondent, CNBC's James Everton, yells out: "Mr. President, is there going to be another attack?"

The president continues to read from his prepared remarks.

"The Department of Transportation has provided more than four hundred trucks to move a thousand truckloads containing five million meals ready to eat, thirteen million liters of water, ten thousand tarps, three million pounds of ice, one hundred and forty generators—"

Something about the triviality of ice in all of this sets the room off. Reporters begin to shout. Katherine Holders of ABC News calls out, "The images coming out of New York City are horrifying, Mr. President. Thousands of people are dying in the streets. What are you doing to help them?"

The president, looking weary, tries to answer.

"Uh, well we have deployed more than fifty disaster medical assistance teams from across the country. Isn't that right Dr. Wasserman?" He turns to look behind him. The Health and Human Services Secretary nods half-heartedly as the president continues, "...to help those in the affected areas."

Holders follows up: "Fifty teams, Mr. President? For all those people? Are they there now?"

"No, they are not there yet. But we do expect that they will begin to arrive within the next seventy-two hours. Once they do arrive, they will be trained in radiation measures and properly equipped to keep them safe."

Somebody hears Holders say, "Too late..."

A cacophony ensues. The president holds up his hands and asks for calm. When it begins to quiet down, a question rings out: "What do you say to people who want to know what they should be doing right now? Who should evacuate and who should stay inside?"

"Well, some should shelter in place and some should evacuate. It depends on where they are. They should listen for directions from local authorities," the president responds.

The reporter follows up: "But people aren't hearing anything. How can you tell people to listen to local officials that you have just said can't be reached?"

The president pauses and begins to answer but is interrupted when someone asks: "What should people do if they are told to stay inside but don't have food or water?"

The president looks down. "We have been in touch with the American Red Cross. They are mobilizing volunteers from across the nation. But they say that their people are not trained to work in those conditions—"

Another reporter asks: "What about here in Washington? Are we out of danger? Many of us can't reach our family members."

From the back of the room, someone shouts, "When will this be under control?"

from bad to worse | *Early October* | *Washington, D.C.*

As the weeks pass, the world watches as the situation descends deeper into chaos.

Politicians in unaffected states begin to call for increasing their insulation from the impacts of the disaster. The Texas governor calls for secession. Within a year, Texas would be an independent country with the former governor as its president. The action meets with condemnation from the president, but there is no direct action from Washington.

Articles of impeachment are drawn up for the president, but many in Congress are objecting, saying it is a futile distraction at this critical time. The president resorts to making speeches meant to rally the country, but that largely fall flat. He ignores continued calls to visit the scene, never venturing farther north than Baltimore, preferring instead to stay a safe distance away.

At the same time, he decides to start a war with Pakorea, almost as an afterthought. He orders a tactical nuclear strike on the People's Navy headquarters. The blast does not kill the Pakorean leader, who thankfully does not respond with more nukes.

The war transitions to a conventional war, and the president spends most of his time leading the war effort from the United States Central Command (CENTCOM) headquarters in Florida. He prefers working with the military because, although they have lots of questions too, unlike FEMA, they have some answers.

The White House takes over FEMA incident communications and begins to issue the situation reports in a glossy daily newsletter format with pictures, highlights, and human-interest stories entitled, "News from the Front Lines."

The newsletter tells stories of heroism and sacrifice that are emerging from the no-go zone of Midtown Manhattan—first from those FDNY and NYPD teams, like 18 Squad, and then from surrounding states that came in before the borders were closed.

Humanitarian relief teams from around the world begin to step into the breach. Even though its headquarters was destroyed, the United Nations, through its Office for the Coordination of Humanitarian Affairs, or OCHA, starts to bring these different teams together into a more coherent response. The OCHA framework allows teams from other countries with rescue capability and who are unafraid of radiation—particularly post-Chernobyl Ukraine and post-Fukushima Japan—to perform life-saving work in the affected areas.

One story that gives much hope involves the Sierra Leone Red Cross Safe and Dignified Burial teams. These teams worked to prepare and bury nearly three thousand bodies during the West African Ebola epidemic in 2014. After the Polaris attack, they entered the blast zone and rescued dozens of survivors, displaying a bravery that made them the first great heroes of the disaster.

As the weeks stretch into months, one thing becomes clear: the mighty USA, the world's lone superpower, has been brought to its knees.

a world gone mad

"Statesmen and generals on both sides of the iron curtain guaranteed to kill everybody there if you kill everybody here; together we preserve humanity by agreeing to eliminate it."[43]

—Lewis Lapham, "Petrified Forest"

Visions of nuclear war haunt those of us who remember the Cold War.

The seeds of the Cold War were planted in Nazi Germany in December 1938, when German physicists[44] discovered fission, a nuclear chain reaction that could produce vast amounts of energy for electric power—or for giant explosions.

The next year, Albert Einstein sent a letter to President Roosevelt warning him that Germany was trying to use this discovery to build a new type of superbomb. Realizing the danger, the president ordered the US government to rush development of its own superbomb. The result was the top-secret Manhattan Project.[45] On July 16, 1945, Manhattan Project scientists gathered at the Trinity test site near Alamogordo, New Mexico, to watch the world's first atomic detonation.

Three weeks later, an American B-29 bomber, the *Enola Gay*, dropped the first atomic bomb on the city of Hiroshima, Japan, killing 210,000 people.

Four years after that, the Soviet Union successfully tested its own device. For the next forty years, the superpowers would be locked in a Cold War, characterized by threats, propaganda, and high-stakes proxy wars. Each built its own arsenal of mighty bombs, along with state-of-the-art systems that it could use to launch them at the other's cities.

> *"The arms race between the United States and the Soviet Union proceeded in earnest, producing decades of constant dread punctuated by deeply terrifying moments of crisis."[46]*
>
> –Jeffrey Lewis, "Our Nuclear Future"

Military strategists justified this perilous situation with a twisted psychological logic: my enemy will not launch all his missiles at me because he knows that I will have enough time to launch all of my missiles back at him. This doctrine—that a full-scale attack would destroy the attacker—was known as mutually assured destruction, or by its more appropriate acronym: MAD.

MAD was no secret.

In fact, the threat of nuclear annihilation was ever present, penetrating the American consciousness and shaping our popular culture. This was perhaps most clearly seen in a phenomenon known as the American teenager:

> "A product-hungry, pleasure-seeking individual, perfectly suited to inhabit a world that could be blown up at any moment and living for the moment with no thought of the tomorrow that might not exist anyway."[47]
>
> —Jeff Nutall, *Bomb Culture*

Some argue that MAD rendered us incapable of conceiving of life with a future.[48] But they are wrong. Americans were concerned about our future, but we planned our lives like everybody else. We just managed the fear in the simplest way possible: by blocking it out. To do that, we needed a wall—rock hard and impenetrable—beyond which was the thought that everything that we knew and loved could be instantaneously incinerated by forces beyond our control. We built a brick wall of hope.

duck and cover

As the nuclear age dawned, the federal government took the lead in preparing the nation for this new and existential threat. In 1950, the National Security Resources Board issued the Blue Book, a 162-page "bible" that guided Civil Defense planning for the next forty years. Civil Defense had several parts, including an education program, evacuation planning, and an emergency alert system. It also included continuity of government planning that led to the first emergency operation centers in the basements of many city halls.

People around in the 1950s and early '60s remember the Civil Defense education program: "There was a turtle by the name of Bert, and Bert the turtle was very alert; when danger threatened him he never got hurt, he knew just what to do... He'd duck! And cover!"[49] And those of us who watched too much television as kids can recite the test script from the Emergency Broadcast System.[50]

But the federal government's heart was never in the job. Complacency plagued nearly every aspect of Civil Defense in the decades after its creation.

Congress never came close to meeting budget requests for a program that bounced between departments and underwent nearly a dozen name changes and agency affiliations before eventually becoming the Federal Emergency Management Agency (FEMA) in 1979 and, after yet another organizational reshuffling, finally ending up with the Department of Homeland Security in 2003.[51]

In the years since, despite increasing concern in the intelligence community, at the White House, and among the public, planning for the nuclear threat has been fractured, haphazard, and ineffective.

looking into the abyss[52]

Forty-seven years after Hiroshima, the Soviet Union collapsed, ending the Cold War. Yet, we live with the legacy of the Cold War even today. A military aide follows the president everywhere with the "football," a briefcase that contains the launch codes for our nuclear arsenal.

Today's world looks very different than it did during the Cold War. Nine countries are known to have nuclear weapons: the United States, Russia, Britain, France, China, Israel, India, Pakistan, and North Korea. Collectively these countries possess more than ten thousand nuclear weapons, most of them many times more destructive than the Polaris. The most common nuclear weapon in the US stockpile, known as the W76, has an explosive force of one hundred thousand kilotons, or one hundred thousand tons of TNT, five times the size of the bombs dropped on Hiroshima and Nagasaki.

The United States and Russia keep their arsenals on alert, able to be launched within the short (thirty-minute) window between when enemy missiles would be detected by our satellites and when they would arrive.

We call this mission "launch under attack," and it leaves almost no time for decision-making. The president would have less than five minutes to decide that a missile alarm was credible and to give an order to retaliate.

At the same time, the risk is growing that one side may try to "decapitate" the other with a precision strike on its leadership or via cyberattack. This creates considerable instability and increases the risk of false alarms.

destruction not assured

Despite these risks, Homeland Security experts worry less about Cold War weaponry and more about bombs of a much lower power (or yield), the impacts of which would be survivable, possibly even manageable.

If Midtown Manhattan were hit by a ten-kiloton blast—
the approximate size of the nuke that North Korea is now
testing—the contaminated "hot zone" of Midtown would
become a ghost town. Beyond a mile or so away, south of
14th Street or north of 53rd Street, people would be largely
safe from most of its effects.

Scientists have identified some easy steps that, if taken
now, could save countless lives if an attack should occur. This
is true whether it involves the kind of high-yield megaton-
nage bombs possessed by longstanding nuclear weapon states
or the much smaller "beginner" nukes like the Polaris.

the biggest problem is between our ears

> *"The unleashed power of the atom has
> changed everything save our modes
> of thinking, and we thus drift toward
> unparalleled catastrophe."*
>
> —Albert Einstein

Another legacy of the Cold War is that brick wall of hope
between our daily lives and our fear of the nuclear threat.
The bad news is that among the greatest deniers of this threat
is our own government. Because they are human, that brick
wall prevents the people we rely on from preparing the nation
for it.

During the Cold War, our preparations were understand-
ably superficial. Why should we prepare to do something
that we will never have to do (since we will all be gone)?[53]

But that was then, and this is now. Back then, we were all sure to die. Today we know that many will die, but almost all will live. Back then it was pointless to prepare for those who were sure to die. Today it is essential that we prepare for the nearly all of us who are sure to survive.

some disasters are not local

"Like politics, all disasters are local."[54]

—Steven T. Ganyard, *The New York Times*, 17 May 2009

The only thing that could possibly be worse than being a victim of a nuclear attack would be to be the person on the hook to fix it. But who is that person? Who "owns" the mother of all disasters?

If we agree that government is accountable to its citizens in a disaster, that person must be a government official.

As we shall see, when talking about disasters, you must know what you mean when you say "government." In addition to federal and state governments,[55] there are thousands of county, city, town, and township governments.

But this is a detail. Like it or not, in the case of the mother of all disasters, accountability will be laid at the feet of the president.

the president's moment of truth

The moment of truth for the president will not be backstage with the satellite pictures on the cellphone, although that will be a very dark moment indeed; rather, it will come later, probably much later. In the Polaris-attack scenario, it came three days later, after he had watched and heard his team "work the job"; after he had continued to get nothing but blank stares in response to probing questions like, "What are we doing now for those desperate New Yorkers?" and, "What are we not doing that has to be done?" and finally, "When will we be doing these things?"

Eventually that "why does it seem like nobody knows what they are doing?" feeling turned from anxiety to fear and, ultimately, to insight. That flash of insight was the moment of truth, when the president realized two things: the first was that the situation was bad, far worse than he had been told, and the second was that he had absolutely no control over it.

Painful indeed is that moment when an elected leader realizes the reality of his predicament. These agencies (his agencies) that he thought owned the problem for him actually don't. And because those agencies never expected to own the problems, they have not developed an ability to seek them out and solve them.

These agencies are, in many ways, worthless to that elected leader in that moment. And that leaves him helpless.

The feeling in the room at the moment of truth is hard to describe. Anxiety and fear leap to mind. And disgust—disgust for the system that has put the elected leader in that position and disgust for the people who hid from him their own fecklessness.

we must get ready

The Polaris-attack scenario gives you something you don't get every day—a glimpse of the future.

The United States and Russia have developed highly alert nuclear postures that are susceptible to false alarms. North

Korea has signaled a strategy to use its nuclear weapons early in any conflict.

The events described are real. They haven't happened, but they will—not here, not today, but somewhere and someday.

When they do happen, the elected leader will have his moment of truth. It is in this moment that he will be filled with regret, mostly for the things that he didn't do.

Often these are mundane things, like mobilizing the nation to prepare for a catastrophic disaster. Although it may sound boring, if done ahead of time, boring things like that can increase our options. They can even save our country.

So, the question remains: "What should we do now?"

There aren't two answers to that question; there can be only one.

We must get ready.

part II | *new york city in the parallel universe*

4 the parallel universe

the nature of catastrophes

"A sudden crash...of the totally unexpected sea upon the ship produces a moment of crisis, a flash-point, one of those brief instants in time when the primal isolation and helplessness of the human condition are revealed."

—Edgar Allan Poe, "MS. Found in a Bottle," 1833

earthquake scenario part 1 | *Middle of the night* | *Your bedroom*

An angry jolt rocks you out of a sound sleep. You open your eyes; it's pitch black and absolutely quiet and still in your bedroom. Suddenly your mattress rises and begins to roll in waves. The maniac who jolted you awake is back, and now he's under your bed, kicking and shoving up as hard and angrily as he can. Noise fills your ears, like thunder, popcorn popping, and fireworks all at the same time. There is a heavy grinding noise too, like brick buildings rubbing together. After a while, the rolling stops and you fall out of bed. The bedroom is shaking now, violently, back and forth, back and forth; you hear lamps and pictures falling off shelves and breaking.

disasters are different

You fell asleep in one world and are waking up in another.

People who have experienced the first minutes of a major disaster report a strange feeling, like a plunge into chaos. Others feel a sense of unreality, like being in a movie or looking down at themselves from above.

That's because everything is different. In this earthquake scenario, you are cast out of your orderly and familiar reality into a new one that is anything but. This new reality is not just a variation on the theme of everyday life. It is not just some fast-moving time—one end of the spectrum, with daily life being the other. The disaster is fundamentally different, alien, and abhorrent....

It is a parallel universe.

One day you will find yourself in that parallel universe. The good news is that you don't have to be a miracle worker to survive.

earthquake scenario part 2 | *Middle of the night* | *Your bedroom*

You manage to stumble into the living room and scramble under your grandmother's sturdy old coffee table. You press your hand up against its bottom, as if to protect your head. You hope it will hold up to the pressure of two stories falling on it. If you were buried under tons of debris, would you ever get rescued?

You lie there in the noise and the shaking for what seems like an eternity. Ten seconds go by, then twenty, then thirty. The longest thirty seconds of your life. You begin to think that this is the end for you. And then, finally, it stops.

doing something is always better than doing nothing

You will be surprised at how quickly your elation at having survived the earthquake begins to fade. Because it will be at that point that the hard realities of a post-earthquake world will begin to filter in.

It's the middle of the night. When you went to bed, your plan was to sleep for a few hours and then get up and go to work. Of course, all of that is gone now—evaporated, out the window.

You were comfortable in that old reality. You had things pretty much figured out. You are not comfortable in this parallel universe, and you'd really like to get back to the other one. But before you can, you're going to have to figure this new reality out.

In this, your moment of truth, you look around and all you see is bad things. *Is that gas I smell? I wonder how my next-door neighbor Mr. Vandross made out? Looks like the power is out. I heard somewhere that I should be filling the bathtub with water. I turned the tap, and nothing is happening. I tried to call Mom, but the cellphone isn't working. Maybe I should turn it off to save power. I would feel better if I knew there was somebody out there who could tell me that I'm not crazy, that this has really happened. Didn't I have an old transistor radio around here?*

The old reality was our modern society, made possible through the benefit of the highly developed human brain. Humans are exceptional because we use that highly developed brain—in combination with our senses—to think and reason, to get organized and solve problems, to build things. Thinking is the key to nearly everything in the old reality. It is also the key to working your way out of this new reality and getting back to that other one.

Your thinking machine has started to shut down, just when you need it the most. The crisis has activated the most ancient and primitive part of your brain. The amygdala, commonly referred to as the lizard brain, is where your primal instinct to fight, take flight, or freeze resides.

In the old days, when your ancestors were lizards, their survival instinct would suppress all other thinking in times like this to focus on fighting or fleeing. This is good when your best option is to run away but bad for nearly everything else.

It's not easy to figure out your next steps when the amygdala is in control, because it destroys your higher-level thinking in favor of panic. It releases a cocktail of hormones, such as dopamine, adrenaline, and cortisol. As the hormone cocktail surges through your body, your muscles tense and you freeze.

Your brain has a very limited capacity for processing new information on a good day. Thinking requires you to hold on to new information while you try to decide about it. But in this disaster scenario, your working memory is busy trying to process all the information that this new reality is throwing at your senses.

Meanwhile, the hormone cocktail is working to shut down your prefrontal cortex, which is responsible for higher-level thinking. So just when you need your wits the most, they abandon you. That is why you will begin to behave strangely.

With your brain shutting down, you will start to rely more on your feelings than on the reality that is before your eyes. You will ignore stressful thoughts and information and reassure yourself by explaining away the danger. You will fall back into your old routines, trying to solve problems in familiar ways, sometimes again and again, regardless of the results. You might just act as though nothing is happening.

But this would not be a good move and is likely to result in suboptimal outcomes. Because in the parallel universe, doing something is always better than doing nothing.

What should you do?

If you are in danger, your best first step is to get out. If you are not in danger, your best first step is to pause, just for a second, to prepare yourself to engage this new reality.[56]

Take a deep breath and hold it for a second. Slowing the breath calms you and walks back some of the damage done by the lizard brain. Feel yourself start to relax; let your shoulders drop and your eyes close. After a minute, start to counter the negative voice in your head with a positive one. Talk to yourself (even out loud). Think of a positive phrase; something like: "*Everything's gonna be alright.*" Keep repeating it to yourself, like a mantra, over and over. Believe that, as bad as it looks now—and no matter how bad things get—you will get back to that old reality.

But you can't get there by sitting back and waiting to be saved. It is time to take ownership of your situation.

Turn the switch in that highly developed brain of yours to *On* and begin to use your mind—in combination with your senses—to act. Because there are a lot of things that, if done now, will improve your situation in the coming hours and days.

The first question is always the same: *should you stay or should you go?*[57] Think about whether it is safe to stay where you are. Become aware of the building you are in—look and listen. Listen for cracking or creaking sounds that signal it is unstable. Look carefully at the inside corners, where the floors, walls, and ceiling meet. If the lines of the corners are tilted or askew, get out.

If you stay, find the gas valve (if you live in an apartment; it's usually behind the stove) and turn it off. Remain alert for that distinctive "rotten egg" smell of gas. Look for smoke, broken pipes, exposed electrical wiring, and other dangers. Be prepared to brace yourself when the aftershocks come.

Find that transistor radio and follow the directions of local authorities. If you hear reports of widespread damage, power blackouts, and boil-water orders, your best course is probably to leave. Listen for the location of the closest public reception center or shelter, and once you are ready, grab your "Go Bag" and get out. Walk down the middle of the street if you can.

If you are staying in your home, fill your bathtub and sinks if you can, making sure to tightly plug the drains. Check for water stored in your water heater (there is a drain valve near its bottom edge) and toilet tank. Keep your refrigerator and freezer doors closed.

Save your phone battery by turning on the power-save mode, shut off push notifications, decrease screen brightness, and shut off GPS, location services, and Bluetooth. Send email or text messages to friends and family telling them where you are and what you plan to do. Write down important phone numbers in case your phone dies. Monitor websites and social media for information.

Understand your situation by finding the decision points: think, choose a course, then act. Over the coming hours and days, you should continue to navigate through the parallel universe using this basic process.

Think. Choose a course. Act.

As hard as it is,[58] and as tired as you'll get, you must make a conscious effort to remain alert and focused and continue to repeat this process over and over. Of course, none of this is obvious and most people will be left trying to figure it out.

In this scenario, thousands, even millions, of other people are impacted in the same way that you are. Like you, their feelings of comfort and sense of order have been destroyed. Like you, they have so many questions to which nobody has any answers. Fear seeps in to replace the destroyed rhythm of daily life. They get a strange sensation that feels like the fabric of society unraveling. Many will do what they always do when faced with big decisions: they will freeze. They will hunker down and try to distract themselves until help comes along and someone tells them what to do, trapped in a parallel universe of suboptimal outcomes.

take a tour of the parallel universe

Unlike our ordered and (sometimes) rational world, in the parallel universe, chaos and confusion reign. The normal rules of logic, even basic ones like cause and effect, don't apply. It is a strange place, so strange that it is difficult to describe. Better to go there and see for yourself.

Because real-world disasters can be messy, we generally do this through simulation. We use our minds to imagine, in as much fine-grained, colorful detail as possible, a disaster scene (or scenario). Let's do that now. We will start with a thought experiment. Think about yourself for a second. What does your calendar look like for the next month, the four weeks starting from today?

Forget about work. Think instead about all the nonwork activities that you would typically do during your free time. Aside from staring at your smart phone, these activities could include things like birthday parties, trips to the beach, reading David Baldacci thrillers, binge-watching *Game of Thrones*, or taking your partner to lunch at that nice restaurant with the outdoor tables on the sidewalk.

Write all those things down.

Now imagine that it's a beautiful fall day and you're thinking about meeting a friend for a drink after work. Just as you hop into your car to go there, your cellphone rings. And, of course, you see that it's *work* calling.

Okay, next pick up that list of yours, all of the fun, meaningful, nonwork stuff you had planned for the next month. Tear it into bits, crumple it, whatever, and throw it in the trash.

You won't be needing it.

Because you are going on a trip. And you are leaving now; leaving your ordered and rational world and passing through a wormhole into the parallel universe that is the disaster—to face the crisis....

active-shooter scenario part 1 | *4 p.m.* |
Friday, October 14 | Downtown

You are driving away from that client meeting that you dreaded so much. It went pretty well; you managed to stay on everybody's good side for a change and your mood is starting to improve. With a long workweek behind and a three-day weekend ahead, good thoughts start to creep into your head. You feel your shoulders relax a bit. It's a beautiful day, clear and crisp, the sun creeping down toward the horizon, with dusk an hour or so away.

A friend mentioned meeting for a drink at that new tavern next door to the Starbucks on Woodmere, so you start to head over...

...and then your cellphone rings.

You see that it's your assistant calling. You pick up but there is so much background noise you can hardly hear her. You hear enough to know something is wrong. Very wrong.

She is distraught. "Oh my god, Megan, where *are* you?"

"What happened, Cara?"

"That guy Bob, Bob Norix I think, the guy you fired last year...he came in with a gun and shot everybody...he got in through the back door and just started shooting.... When I heard it, I ran out the front door and down the street... and then the police came, and they killed him.... He shot Emily and Alex...they're dead, I think they're dead.... Clay and Ryan are shot, too, and I don't know how they are..."

You quit a high-paying job in Silicon Valley seven years ago to strike out on your own. With thirty-plus staffers and three offices around the city, your boutique tech firm is thriving. You had no idea when you started that you were on the hook for something like this.

You feel the panic start to build; right now, you know you can't let it overtake you. This company is your baby. You own it, and you have no choice but to jump into the fray with both feet.

Somebody held open a locked rear door to let a gunman into your workplace. Four employees are killed; four are severely injured. The shooter, a disgruntled former employee, is dead.

You try to make phone calls in your car on the way back to the office but come to your senses after nearly running a red light. You focus instead on gathering your thoughts. "What," you start to think, "are the issues I need to deal with right now? Four of my staff are dead, at least four more are critically injured, the office is damaged, staff are still there, and there may even be clients there. On top of that, the police will want to talk to me. I have to talk to my lawyer, my investors, my insurance agent, and my landlord."

So that's a dozen issues, some big, some medium size, some huge. Your firm is a tech project management firm, so you try to project manage the situation. The first step is to figure out which of these issues can wait. Nothing is leaping to mind.

An image of a smiling Emily and Alex interrupts your thoughts. They are squinting into your cellphone camera at the beach. You were with them there, just this past summer. They were more than employees; they were good friends. And now they are gone. Sadness washes over you, from the top of your head all the way down to the tips of your toes.

Police cars and fire trucks are blocking the street when you finally arrive. You can't even get near the building, so you park two blocks away and run into the crowd. The dead and injured are gone, and you try to focus on your staff. The intensity escalates, and you spend the next few hours in a blur, answering questions, hugging crying employees, talking, talking, and talking some more.

You are hijacked by the police and keep getting pulled away to talk to the detectives, often for long stretches of time. The people you most need to talk to are your key employees. They experienced the trauma of the event and of losing their friends and coworkers.

Your assistant seems to have the best handle on the situation and is connected to most of the people who are there, all of whom want to talk to you. Right now. She is looking at you while half a dozen people in a line are looking at her.

the "spectrous fiend"[59] that is the crisis

It could be anything—widespread and all-encompassing, like a natural disaster, or focused and devastating, like an active shooter. Whatever kind or variety of crisis you find in the parallel universe, it's useful to understand what the crisis is not.

The crisis is not an incident. It is not a concept, not inanimate. It is not random, and it is not benign. It is a being, inhuman and flawed, with thoughts and intentions. None of those thoughts are good, and all of those intentions are bad.

The crisis is a dragon that must be slayed—because it wants to destroy you. It puts you into its crosshairs the instant you enter the parallel universe. It knows that when you get there, you will need to think,[60] so it starts with sabotage.

It sends an electric charge through your body, triggering your lizard brain and the hormone cocktail that shuts down your prefrontal cortex. As your higher-level thinking slows, the crisis floods your brain with agonizing thoughts and painful emotions. It takes over the voice in your head: "You can't handle this," the crisis says. "No way can you figure this one out." Or, "Life as you know it is over and it's never coming back."[61]

Leonard Marcus and Barry Dorn at Harvard call this amygdala-controlled state the "emotional basement."[62]

As you stumble around in your dark and musty emotional basement, you are overcome with a jumble of negative emotions: grief, fear, anger, regret, hate…. You grieve for so many things, starting with lost loved ones or friends and extending all the way to lost birthday parties and trips to the beach.

You fear the ordeal. You know you will be hit with one painful problem after another. Some you will be able to solve, but not all—at least not immediately. These will burn white hot in your lap until they are resolved.

You fear the work, because you know that you won't be working like you worked yesterday or had planned to work tomorrow. You will be working twice as hard, possibly harder than you have ever worked in your life (including in law school).

You feel anger and regret for things you did or didn't do that supposedly put you into this situation (for instance, "I should have called the police about Bob Norix" or, "I should have done more to secure that back door").

You hate your life. You hate the world and everything in it.

This is your moment of truth.

This is the fateful moment when you must decide to engage or to disengage. Do you step down and fade into the background? Or do you step up?

Even though the crisis has transformed your brain into mush—has turned your orderly and intelligent thoughts into white noise—you know there is no turning back, no running away from this. You steel yourself, set your jaw, and take that first step; in the desperate hope that step one will eventually lead to step two and maybe even to steps three, four, and five.

In the absence of complete information, in the presence of dozens of unanswered questions, in the face of seemingly insurmountable obstacles, in the presence of danger and risk, you turn and face the dragon.

murphy's addendum

In the run up to that phone call in the car, in the weeks and months before that fateful Friday afternoon, you were confident. You really didn't give it much thought, but had you been asked, you would have agreed that someday a crisis would come and that you would have to face it. You were pretty sure that you were up to the challenge. These things happen in this life, you figured, and you had been in tough spots before. You were confident that with a little luck and a lot of hard work, you would get through it.

Once you were inside the parallel universe, that confidence was shattered—when those things that always worked for you stopped working and the people who were always there for you were gone.

But the crisis did not stop there. It took aim at your most important asset. It messed with your time. It slowed things down and then, in those first moments, brought things almost to a standstill. Then, with a haughty laugh, it cranked the dial and suddenly time began to race: days became hours; hours became seconds.

And all the while it conspired against you, contradicting your every attempt to figure it out. Just when you are thinking A, the crisis throws out Z.[63]

you can't get big enough fast enough

The crisis is deliberate and methodical. It bides its time, gathering strength and power, and building its unique chaotic mix. And, at just the right moment, it unleashes its most fearsome weapon, an avalanche of impacts we call surge. Everything happens at once—everything, that is, except what we want to happen.

active-shooter scenario part 2 | *6:24 p.m.* |
Friday, October 14 | Main office

And then the families arrive...

They are the spouses and mothers and siblings of the vibrant human beings who have been suddenly ripped from their lives. They are overcome with grief.

And they have needs. Mothers collapse on the floor, caring about nothing. They know it is true, but their minds play cruel tricks: "It's a mistake," says the voice in their heads. "She is okay. You will get her back." This is the worst sort of agony one can experience in this life. Many will not eat or sleep. They will need everything—hotel rooms, transportation, clean clothes. They must get constant attention and emotional support. Most of all, they want information.

Grief, a great beast, is pressing down on them and you and everyone around you. You have to figure out how to contain it. You need a quiet space to engage the families and address their unquenchable thirst for answers: "Where is my loved one now? What happened and how did they die? What were their final moments like? Who can I blame?" Only a continuous stream of information will help them, and no detail is too small. You are finding out the hard way how tough it is to support a grieving family. And this is just one of a dozen issues sitting in your lap.

You spend the next several hours at the center of a maelstrom that shows no sign of abating. You got that call in the car at four o'clock; you just looked at your watch and its 11:49. Eight hours passed by in a New York minute.

You still don't know what you are going to do with your office. Are you going to clean up and just open as normal? How are you going to talk to your staff? When are you going to do this? What are you going to tell them? You need a plan.

"I'm not leaving," you think to yourself. "I can get through this." You decide that you will work through the night. When the sun comes up, you're going to feel like a vampire. You'll be exhausted. The maelstrom will be there still.

How did you ever expect that you could get through something like this? You have a dozen big problems and one assistant to help you. Even if you had a dozen people to help you, it wouldn't be enough—because you don't need a dozen people; you need a dozen teams, some big, some medium size, some huge....

extreme multitasking

Although there can be no denying its tragic impacts, this devastating scenario falls on the low side of the disaster magnitude scale. You were the boss in that scenario. Think about how overwhelmed you would be dealing with the issues brought by that crisis.

Now think about what it might be like to be a boss for a more widespread and all-encompassing disaster—like the mayor of New Orleans during Hurricane Katrina, for instance. Through the long, hot days of late August, the crisis had been biding its time, waiting patiently to hit New Orleans with a torrent of chaos. What do you think that surge in human needs looked like on the Gulf Coast on August 29, 2005?[64] How many issues did the crisis bring on that day—twenty? Two hundred? Twelve hundred? How overwhelmed do you think you would be dealing with all of them?

Unfortunately for the affected community, in those early hours, the response community[65] was essentially flat-footed.

Its ability to to meet the need across the entire affected area—to put response teams into the field, pump out the water, get the power back, and clear the streets—was days, in some cases weeks, away.

This unhappy condition is little changed today. We call it the "speed to scale" challenge, and it remains government's biggest problem. For people in the disaster business, such as myself, it means failure and humiliation. For the affected community—the seniors, individuals with disabilities, children, and families—it means suffering on a grand scale.

The crisis understands this. Its surge is a chaotic mix specifically designed to disrupt critical infrastructure—power, water, food, medicine, sanitation, housing, you name it—in unique and unprecedented ways.

To counter this, it's not enough for us to do a lot of things at the same time. Human needs cannot wait, so we have to be able to do everything all at once.[66]

Because the crisis surges, we must surge too. We need to get big enough fast enough to manage a tsunami of issues in its early hours.

Let's be clear about something: Surging doesn't mean doubling down on Starbucks so that everybody can work harder and faster. It doesn't mean calling in your crisis team and rolling up your sleeves, so that with a little luck you will get through it. Surging means getting fifteen times more people on your team than you have now. It means getting fifteen times more resources than you currently have on hand. It means turning your crisis team into fifteen crisis teams, each with a different mission, focused on a different human need.

Are you starting to get the picture?

Now some will disagree with me on this point. They will say "Hold on, let's not go off all half-cocked here. We need to gather first and get organized. Until we get a 'damage assessment' we can't really do much of anything. It's not prudent to move a lot of people or equipment or to spend a lot of money before we get a clear understanding of the situation on the ground."

The people who say these things are under the spell of the crisis (you will hear many of these kinds of excuses inside the parallel universe). If you should ever find yourself within earshot of this kind of talk, this is what you should say: "As a matter of fact, what we need to do now is get as big as we can as fast as we can. Not only do we need to get a lot of people and equipment into the field, we need to reach back and get everybody we know, and everything we can get our hands on, on its way to us."

know the enemy

> *"If you know neither the enemy nor yourself, you will succumb in every battle."*[67]
>
> —Sun Tzu, *The Art of War*

Remember that strange feeling you get in the first minutes of a disaster? The sense of unreality, like a plunge into chaos? Even the people who have been there many times get that feeling[68]…every time. And, after climbing out of the emotional basement a few times, you learn a few things—things like what to do and, more important, how to think, especially about how we must consider that the crisis is our enemy. And the more we know about that enemy, the better things will go for everyone. For instance, we learn that the crisis does not want to be comprehended, so it hides from you. That is why even when the crisis is upon you, you will not recognize it.[69]

The crisis shows a different side of itself to everybody, so there is never one story of the disaster; there are always as many stories as there are people who were there.

Each crisis is an aberration, unique in nearly every way. You will see things that you have never seen before and hear things that you never thought you would hear. Many of these things will require improvisation, new ideas, and novel approaches to resolve.

The crisis lies, so nearly everything you hear, especially in the first minutes, will be wrong.

You think that you're ready for the crisis. But when it comes, it will wipe your mind clean and you will forget everything.

The crisis holds its victims hostage. It doesn't let you talk to them, so that when people in the normal world try to talk through the wormhole to the people affected by it—seniors, children, and families—the most common outcome is outrage.

A good example of this is the Christmas Blizzard of 2010, the storm that came out of nowhere. On the Friday before a long holiday weekend, the National Weather Service predicted five inches of snow for New York City. At midnight, the forecast was increased to eight inches. But by the following afternoon, on Christmas Day, the National Weather Service issued a Blizzard Warning and predicted fourteen inches of snow for New York.

christmas blizzard

And then it came. On the day after Christmas, New Yorkers saw one of the biggest snowstorms in New York history bring transit to a halt and strand ambulances in four-foot drifts. The 2010 Christmas Blizzard dumped between eighteen and twenty-four inches of snow—twenty-nine inches in parts of Staten Island—and brought wind gusts of up to sixty miles per hour. The snow started Sunday morning, on one of the busiest travel days of the year. It fell throughout the day and through the long night.

After midnight on Monday, December 27, it really started to come down. Snowfall rates accelerated to between two and three inches per hour, bringing whiteout conditions. The city began to get reports of snowbound ambulances and buses, and of abandoned cars blocking the streets.

The problem got worse throughout the night. Tow trucks and New York City Fire Department trucks sent to rescue stuck ambulances also got stuck. As the sun came up on Monday, complaints were flooding in from all over the city about snow-clogged roads and missing snowplows.

That morning, Mayor Michael Bloomberg gathered with his leadership team before a hastily called press conference inside a cavernous Department of Sanitation garage in downtown Manhattan.

The Press Secretary and agency commissioners normally spend a few minutes walking the Mayor through a list of talking points before a press conference. But this was not a normal day. Everybody had been working the job since Saturday night and they were exhausted. The Mayor was crankier than usual, and they were pressed for time. They did not push hard enough to get him properly briefed, so the Mayor didn't understand the magnitude of the crisis as he stepped up to the podium.

Bundled in an overcoat and scarf, the Mayor began: "*Yesterday's blizzard was historic, but we are making progress. I ask all New Yorkers to be patient...*" After delivering his prepared remarks, the Mayor opened the floor to questions.

"*The storm has sent city emergency services into a nose dive. Ambulances and fire trucks are trapped in snow and facing long delays,*" said Marcia Kramer, the venerable chief political correspondent for WCBS-TV. She then asked the Mayor, "*Are people dying in the streets?*"

"*People are dying, just, naturally*," Bloomberg replied. "*Just because an ambulance gets there, doesn't mean you can save a person. The science isn't that good.*"

Standing at the podium with his commissioners arrayed behind him, the Mayor grew defensive. Instead of explaining the challenges the City faced in responding to a fierce snowstorm; instead of describing in detail how bad it was in the streets; instead of talking about what the city had not yet done and when it was going to do these things, the Mayor seemed to point the finger at drama queen New Yorkers. "*The world has not come to an end*," he said. "*The city's going on. Many people are taking the day off. There's no reason for anybody to panic.*"

He followed this with advice for New Yorkers on what they should do instead of bitching about snow. "*Broadway is open; the NYC ballet's performance of the Nutcracker is on. People should have gone to the park and enjoyed this time with their families.*"

This struck a chord. New Yorkers were dying in ambulances mired in city streets and the Mayor was telling people to go to the Nutcracker. Reporters who had been leaning against the wall or slumped in their seats suddenly perked up. One pulled a pencil from behind his ear and began to write furiously in his notebook.

The blood was in the water, in the form of outrage, shame, and political retribution, from which the Bloomberg administration would take months to recover.

the quickest route to outrage is to deny the reality

People affected by disaster are trapped in a parallel universe, and you are not. They rightly expect that you are moving mountains on their behalf, but so far the only mountain they see is the mountain of problems that stands right in front of them. They are deeply skeptical of all communication from people, like you, who are on the outside. Everything you say is garbled as it passes through the wormhole and resonates terribly inside their parallel universe.

They see all sides of every issue and will jump onto the side of the issue that you ignore. For instance, too much reassurance will backfire. The more you say 'it's safe' the more they will be convinced it's not. The more you emphasize that it is not your fault, the more they will know it is. The more confident you try to sound, the less confidence they will have in you.

The only way to make any headway with them is to commit to them via a trip through the wormhole. Only when you are physically in their midst will they start to listen with an open mind. Even then you must be very careful. You must always and everywhere convey an intimate understanding of their situation. Every message must include a detailed picture of what they are seeing, feeling, thinking, and experiencing. This is how you build a bridge to the people who matter most. This is how you battle the outrage to create a bond; a bond built on credibility and trust. That's the only way you can get it. There aren't two ways.

what will we be saying after the next really big one?[70]

> *"Subduction-zone earthquakes operate on the...[principle that] one enormous problem causes many other enormous problems."*
>
> —Kathryn Schulz, "The Really Big One"

In her Pulitzer Prize–winning article "The Really Big One," *New Yorker* staff writer Kathryn Shultz tells the story of the worst natural disaster in the history of North America. According to scientists, on or about January 26, 1700, a massive earthquake in the Pacific Northwest ripped a gash in the earth's crust along a line from Vancouver Island in Canada south nearly six hundred miles into Northern California, causing massive devastation. The geological record indicates that these "great earthquakes" (those with a magnitude of eight or higher) occur in this area of the Pacific Northwest about every five hundred years on average.

In "The Really Big One," Shultz describes for us the implications of this revelation. When it comes, the next Really Big One could impact an area of 140,000 square miles and devastate major population centers like Seattle and Tacoma in Washington, and Portland, Eugene, and Salem in Oregon. Seven million people could be cast into this parallel universe, of which nearly 13,000 people could die and another 27,000 could be injured. When it happens, we would need to provide shelter for a million displaced people, and food and water for another two and a half million.

The following excerpt will take your breath away:

When the Cascadia earthquake begins, there will be...a cacophony of barking dogs and a long, suspended, what-was-that moment before the surface waves arrive....

Soon after that shaking begins, the electrical grid will fail, likely everywhere west of the Cascades and possibly well beyond. If it happens at night, the ensuing catastrophe will unfold in darkness.... Anything indoors and unsecured will lurch across the floor or come crashing down: bookshelves, lamps, computers, canisters of flour in the pantry. Refrigerators will walk out of kitchens, unplugging themselves and toppling over. Water heaters will fall and smash interior gas lines. Houses that are not bolted to their foundations will slide off....

Other, larger structures will also start to fail... across the region, something on the order of a million buildings will collapse....

The shaking from the Cascadia quake will set off landslides throughout the region—up to thirty thousand of them in Seattle alone.... It will also induce a process called liquefaction, whereby seemingly solid ground starts behaving like a liquid, to the detriment of anything on top of it.... Together, the sloshing, sliding, and shaking will trigger fires, flooding, pipe failures, dam breaches, and hazardous-material spills. Any one of these second-order disasters could swamp the original earthquake in terms of cost, damage, or casualties—and one of them definitely will.

Four to six minutes after the dogs start barking, the shaking will subside. For another few minutes, the region, upended, will continue to fall apart on its own.

Then the wave will arrive, and the real destruction will begin.[71]

The odds of a big Cascadia earthquake happening in the next fifty years are roughly one in three. The odds of the next Really Big One are roughly one in ten. Even those numbers do not fully reflect the danger—or, more to the point, how unprepared the Pacific Northwest is to face it.

We should pause for moment to take all of this in. The enormity of this breathtaking scenario makes it difficult to contemplate fully. But contemplate, we must.

And then, after we have contemplated for a while, somebody needs to get to work. I have an idea: how about we build a Pacific Northwest Cascadia Subduction Earthquake and Tsunami Response Plan? The PNCSETRP (as I like to call it) would be massive and unprecedented, nothing less than a comprehensive, proactive, integrated, and all-of-nation plan. Although it sounds complicated, all you really need to do is to put all the people who would be responsible for a Pacific Northwest Cascadia subduction earthquake and tsunami response in the middle of an imagined Cascadia subduction earthquake and tsunami to figure things out ahead of time, instead of in the fog of war. Thanks to Kathryn Shultz's elegant scientific narrative we have an incredibly detailed imagined disaster to work with.

So, let's do that now. Let's imagine that it's 2:35 p.m. on a rainy Saturday afternoon in March and the next Really Big One hits.

We need to think through exactly what that would look like. We need to quantify the unprecedented surge that the crisis will bring. We need to understand, in as much fine-grained, colorful detail as possible, that enormous problem that causes so many other enormous problems. We need to list all of the issues that we—the United States and the world—would be dealing with as that Saturday afternoon turns into a long Saturday night. We need to think about the people—the seniors, the individuals with disabilities, the children and families—who would be trapped inside that parallel universe.

Instead of trying to think through these things then, we need to do it now, so that we know what we will tell them about when we are going to reach them. About how we are working across 140,000 square miles of affected area to rescue people from collapsed buildings, pump out the water, get power and cell phone service back, and clear the streets. About how we are providing shelter for a million displaced people, and food and water for another two and a half million.

To be able to do these things then, we need to get to work *now*.

We must travel through the wormhole and into that parallel universe, to spend as much time as possible in the Pacific Northwest on that Saturday afternoon with those collapsed buildings, blocked roadways, stuck trains, trapped victims, dead and injured people, and debris in the streets. We must figure out everything we would have to do all at the same time, who is going to do it, and where we are going to get all of the stuff we will need to make it happen.

I hate to be the one to break it to you, but we are not doing this work today.

Instead, in cities and states all over the Pacific Northwest, and the nation, disaster professionals sit around in small groups in carpeted conference rooms, using rational thought processes to write pieces of the plans about pieces of the job they think they own. And, by the way, these plans have been shown to work spectacularly well...in carpeted conference rooms.

There is no substitute for an integrated, all-of-nation planning process like the one described above. So why are we not doing it? Why do we instead sit in carpeted conference rooms with our cliques telling war stories, asking the same old questions, and speaking the same tired platitudes? Why, instead of spending time trying to understand the enemy, do we clutter our minds with process and unrealistic expectations—so that we are surprised, caught off guard, when the realities that the crisis inevitably brings don't fit our processes or expectations?

Why indeed.

5 the myth of the national disaster system

we are not prepared

"The first responsibility of government is to preserve the public order."[72]

—Robert Ellsworth Wise, Jr.

katrina aftermath part 1 | *August 30, 2005 | Emergency operations center | Downtown New Orleans, Louisiana*

Karl Metairie, New Orleans' deputy emergency preparedness director, has been awake for more than forty hours. As he navigates through the chaos of his dimly lit Emergency Operations Center on the ninth floor of the city's chief administrative offices, he doesn't see a nap anywhere in his future. On a typical day, thirty people could comfortably work in this space, the size of a large conference room. There are at least a hundred people here now, standing in groups and sitting at desks, some calm, some not so calm, some in conversation, some on the phone, some staring blankly at a bank of television monitors mounted on the walls.

Overnight he had ordered his logistics team to move downstairs to a conference room on the eighth floor. Nobody has heard from them since. The ring of his two cellphones mixes with the ringing of the desk phones as talking heads and gruesome scenes flash across the television screens.

Metairie needs to talk to his boss, but some US Army general and a hospital administrator are trading elbows to talk to him. There are lines four or five people deep behind them. The city's press secretary pushes his way through the crowd and waves at him. Metairie owes him talking points for the mayor's press conference that starts in five minutes.

A couple of hours ago, Metairie spent half a minute trying to come up with those talking points. But there is so much he doesn't yet know. He knows that nearly every levee in metro New Orleans has been breached, and that most of the major roads in the city are damaged, including the I-10 bridge. Reports are starting to come in about people showing up at the Superdome (the "shelter of last resort") and the convention center, with no place else to go. Looking out the window, he can see the blownout windows of the Hyatt Regency, with some beds hanging out of them. The city and the world need to hear from the mayor. But from inside the chaos, even Metairie doesn't even know how bad it is out there.

The local EOC is the epicenter of every catastrophe, and in those early days after Hurricane Katrina made landfall, the New Orleans EOC was the busiest place on earth.

Through the long, hot days of late August, the crisis had been biding its time, gathering strength to hit New Orleans with a chaotic mix intended to disrupt critical infrastructure—power, levees, roadways, food, medical and sanitation services, housing, you name it—in unique and unprecedented ways.

Karl Metairie and his team struggled with a multitude of issues—twenty? two hundred? twelve hundred?—in those early days of the crisis. And they needed help.

But where was that help? Where was the cavalry when New Orleans needed it?

katrina aftermath part 2 | *August 30, 2005* | *FEMA command post* | *Shreveport, Louisiana*

Meanwhile, three hundred miles due north, in Shreveport, Melanie Barton sits in quiet, air-conditioned comfort. A young FEMA staffer, Barton was told that she would be assigned to a national emergency response team and sent to Louisiana. Three days ago, she got on a plane and flew to Baton Rouge.

Barton wasn't sure what she was supposed to be doing. Like the nine or so other young FEMA staffers on the Blue Team, she hadn't been trained for this; she didn't even know whom to go to with questions. They were busy when the Blue Team arrived at the State of Louisiana EOC in Baton Rouge—besides, there was no room for them there. Someone told them to establish a "command post" instead. So here she sits, in a hotel ballroom in Shreveport, watching the horror unfold on the television as her colleagues chat with friends on the phone or casually surf the internet.

The calm of Barton's FEMA "command post" contrasts starkly with the chaos at the New Orleans EOC. Although FEMA had deployed its emergency response team to the state of Louisiana's EOC before Katrina made landfall, it was unable to combine forces with the state. In a similar way, the state was unable to combine forces with the city of New Orleans.

In those crucial early hours, when the government-led response needed to get big enough fast enough to manage the surge in human needs that the crisis brought, its various pieces were all but disconnected from one another, working at cross-purposes—and losing time.

Unfortunately for the affected community—the seniors, individuals with disabilities, children, and families—the government[73] was flat-footed. Help would not arrive for days, in some cases, weeks.

the responsibility of government during emergencies

Nine out of ten Americans say that government should play a major role in responding to disasters.[74] That is because we rely on government for the services that only governments can provide. These services—firefighting, law enforcement, and national defense, for instance—bind our social fabric and make our ordered daily world possible. This is the most solemn obligation of government—to preserve the public order.

Preserving order means rushing to the scene of emergencies, from car crashes to house fires to hurricanes. It means fixing damaged critical infrastructure and helping people while at the same time providing all the other services that are needed every day in a modern functioning society—things like fixing potholes and teaching schoolchildren.

In the disaster business, we say that government "owns" the disaster.

The term "ownership" means different things to different people. For some, it means owning things, like a house or car, while others point to the concept of taking personal ownership of one's choices in life. During disasters, ownership is more like that second one. It means accountability. It means taking the initiative even without full authority for the process or the outcomes. This kind of ownership goes beyond mere responsibility. Responsibility is delegated by a boss or by virtue of a position. Those who own the disaster must be willing to face the consequences that come with failure.

This is what government does. When everybody else runs away from the disaster, it must run at it. It is ultimately responsible for it and accountable for what gets done and doesn't get done. That is the end of the story.

Except when it isn't.

what exactly do you mean when you say "government"?

When trying to figure out who owns the disaster, you first must know what you mean when you say "government." Contrary to what some people think, we don't have just one government in the US.

In addition to our three levels of government—federal, state, and local—[75] there are thousands of county, city, town, and township governments—over 36,000 in all.

Because, as the name implies, we are a republic of states, it is these individual states that serve as our primary unit of government. And through the US Constitution, the sovereign states agreed to cede certain powers and responsibilities to the federal government.[76] These powers and responsibilities, however, did not include disaster response.[77]

Since the responsibility for disasters was not assigned to the federal government, ownership must therefore reside with the states. But, as it turns out, the states don't own the disaster either.

Home rule is a type of law that delegates power from the state to its subunits of government.[78] Through home rule[79] laws, most states[80] have transferred the responsibility for disaster response to the government that is closest to the people—to the counties, cities, and towns. The states therefore have pushed their ownership stake down to where it currently resides: with local government.

> *"Incidents begin and end locally."*
>
> —FEMA Administrator Brock Long [81]

The idea that all disasters are local refers to the fact that local governments—some thirty thousand across the country—bear the primary responsibility for disasters. Your local elected official is responsible for knowing what is going on in your neighborhood, especially with respect to its most vulnerable citizens—the poor, the sick, seniors, and the disabled. That official is expected to work tirelessly on their behalf, to advocate for them, to identify and address their unmet needs.

That responsibility does not change when a disaster strikes. Indeed, it becomes even more vital. After the next disaster, you and your family could be left to fend for yourselves, when the power has been out for days and the cell phones are dead and downed trees are blocking the street to your house.

After you haven't seen anybody—not the power company or the health department or even the mail carrier—you will begin to wonder, "Who is in charge?" The answer is, the same person who is in charge today: your local elected official—your mayor, county executive, alderman or alderwoman, or county judge.

This is our disaster response system, and it works well... except when it doesn't. We don't have to go too far back to see examples of this.

victory lap

> *"In Puerto Rico, the 'natural disaster' is the US government."*[82]
>
> —Yxta Maya Murray

The hyperactive 2017 hurricane season featured six major hurricanes, the most since the devastating 2005 season.[83] It was the costliest hurricane season on record, with more than 281 billion dollars in damages, nearly all of which was due to the three major hurricanes—Harvey, Irma, and Maria—that made landfall in the United States.

It was also unique in terms of how we responded. By most accounts, the government-led disaster response for the 2017 hurricane season was a home run. After Hurricane Harvey, based in Texas, and Hurricane Irma, based in Florida, *The New York Times* signaled that it was time for a victory lap, saying that "government disaster response has grown more sophisticated."[84] And then Maria showed up.

Hurricane Maria made landfall on the commonwealth of Puerto Rico on Wednesday, September 20, as a Category 4 hurricane with 115-mile-per-hour wind gusts and more than two feet of rain. That day, the governor of Puerto Rico announced that the island was "destroyed," with 100 percent of the power grid knocked out. Nearly a month later, less than fifteen percent of Puerto Ricans had electricity, more than thirty percent had no access to drinking water, and nearly half had no cellphone service.

Many public figures were critical of the government-led response, including San Juan mayor Carmen Yulín Cruz, who called the disaster a "terrifying humanitarian crisis" and pleaded for FEMA to do more. Frustrated with the federal government's "slow and inadequate response," Oxfam America president Abby Maxman said, "…the situation in Puerto Rico worsens and the federal government's response continues to falter…"[85]

How can we account for what *USA Today* calls the "wildly different responses to the hurricanes in Texas, Florida, and Puerto Rico"?[86] Why does our system seem to work well for most disasters, but not for all?

some disasters break the system

Let's start by looking more closely at the parallel universe that we call "disaster."

The formal definition of "disaster", according to the World Health Organization, is "a serious disruption, occurring over a relatively short time, of the functioning of a community or a society involving widespread human, material, economic, or environmental loss and impacts." The other important aspect of disasters is that their impacts "exceed the ability of the affected community to respond effectively."[87]

But disasters whose impacts exceed the ability of the affected community come in all shapes and sizes, from plane crashes to human extinction, and categorizing them is notoriously difficult. For the purposes of this discussion, we will define two categories of disaster: *home-rule* and *Maria-class*.

We will say that a home-rule disaster is one with impacts that exceed the ability of the affected community but are within the combined capabilities of state and local governments to respond effectively. And we will define a Maria-class disaster as one with impacts that exceed the combined ability of state[88] and local governments to respond.

Fortunately for us, most disasters in the United States are home-rule disasters. Also, fortunately for us, the government is getting better at responding to these.

"The one thing I just want to repeat is how proud I am of FEMA.... [Craig Fugate] has done such an outstanding job [and] has really rebuilt [it]."[89]

—Barack Obama, August 23, 2016

The "bromance" between President Obama and his FEMA director was one of the Beltway's worst-kept secrets ("I love me some Craig Fugate," the president once said[90]). In August 2016, when heavy rain caused flooding that submerged thousands of homes and businesses in Louisiana, FEMA got on the ground quickly and delivered food, water, and medical care and conducted search-and-rescue operations.

Now, when talking to FEMA about its work in Louisiana, you have to refer to its DR (for disaster response) for the incident entitled "4277-Louisiana Severe Storms and Flooding." DR-4277 is one example of a good job done by FEMA over the past several years; others include the Hurricane Harvey and Irma responses in 2017, the Texas floods and Hurricane Matthew in 2016, Hurricane Sandy in 2012, and the Joplin tornado in 2011.

DR-4277 was what FEMA calls a major disaster. Major disasters are defined by Congress as "any natural event, including any hurricane, tornado, storm, high water, wind-driven water, tidal wave, tsunami, earthquake, volcanic eruption, landslide, mudslide, snowstorm, or drought, or, regardless of cause, fire, flood, or explosion, that the President believes has caused damage of such severity that it is *beyond the combined capabilities of state and local governments to respond*."[91] In the first ten months of 2017, President Trump signed nearly 125 major disaster declarations. There were more than 1,200 major disasters in the United States in the decade leading up to 2016.

Even though they have the same definition, Maria-class disasters and FEMA's major disasters are quite different. That difference, between a true Maria-class disaster and a FEMA major disaster, is the key to understanding the disasters that break the system.

Because, as bad as it was, the impacts caused by the flooding that Louisiana experienced in August 2016 were *not* beyond the combined capabilities of the State of Louisiana and its local governments. Notwithstanding this, FEMA approved a request from the governor and a major disaster was declared by the president on August 14, 2016.

The same thing happened during Hurricane Harvey and Hurricane Irma, the Texas floods, Hurricane Matthew, Hurricane Sandy, and the Joplin tornado. As bad as these disasters were, and they were all bad, none exceeded the capabilities of the affected states.[92]

Despite this minor technicality, *all* were declared major disasters by the president. This happens time and time again. In fact, of the more than 1,200 events declared major disasters by the president in the decade leading up to 2016, not one was beyond the combined capabilities of state and local governments to respond.

Many will argue this point. They will say that federal support was important to the response, that the food, water, and ice that FEMA brings is good for a state during a disaster response.

It's hard to argue with that, but should you ask the governor whether her or his government was helpless and overwhelmed, the answer will be not just "no" but "hell, no." If that governor is honest, she or he will admit to wanting a presidential declaration for one reason and one reason only: free money.[93, 94]

there ain't no such thing as a free lunch

> *"Trump says states can count on federal cash in emergencies."*
>
> —Associated Press, August 4, 2017

When the president declares a major disaster, seventy-five percent of all response costs are covered by the federal government. Even better, the state's twenty-five percent share is routinely waived. FEMA picked up 100 percent of response costs after 9/11, for instance. For the Hurricane Harvey response in Texas, the president authorized a 100 percent federal cost share for what FEMA calls "emergency protective measures."

"What's wrong with that?" you may be thinking. "Nothing is better than free money, right?"

First, as it is with lunch, there is no such thing as "free money" (more about this later). Second, and more important, the expectation that the federal major disaster declaration process creates is a problem. That problem is an unfortunate phenomenon known as moral hazard.

"Moral hazard" can be defined as "a lack of incentive to prepare for disasters when one believes he or she is protected from their consequences." The moral hazard created by the federal major disaster declaration process causes state governments to underprepare for Maria-class disasters. Instead, they fall under the spell of the "cavalry is coming" myth.

This myth tells them that they don't need to work that hard to prepare for disasters because all the support they will ever need is there for the asking. The "cavalry is coming" myth has succeeded in turning a fiction—that a massive federal response can be triggered merely through a request for support—into conventional wisdom.

the hazards of moral hazard

As an example of this, we have to look for a Maria-class disaster. Fortunately for us, true Maria-class disasters are rare. Other than the failed Hurricane Katrina response in 2005, we must go all the way back to 1992, when a Florida politician named Lawton Chiles was less than a year into his first term as governor.

On Monday, August 24, Andrew, a Category 5 hurricane, plowed through the town of Homestead, Florida, completely obliterating the houses and stripping them off their foundations. In Miami-Dade County, the 165-mile-per-hour winds destroyed more than 25,500 houses and damaged more than 101,000 others, leaving sixty-five people dead along its trail of destruction. And then came the aftermath…

"WASTELAND—Across the region, the looting that had begun as pure plunder became more an act of desperation."—Newsday

"CATASTROPHIC —As we moved into the response mode, things kept getting worse. We were faced with hundreds of people who needed emergency care with no place to take them."—The Times-Picayune

"HOW DID THIS HAPPEN?—The air reeked with dead animals, human waste and soaked fabric. A message was sent to the President that the National Guard could not control the mob, feed the hungry, rebuild the infrastructure and police the streets."—TIME[95]

Three days after Andrew made landfall, Dade County emergency management director Kate Hale, during a televised news conference, said: "Where in the hell is the cavalry on this one? They keep saying we're going to get supplies. For God's sake, where are they?"[96]

The people of Dade County had been left largely to fend for themselves up to that point.[97] In the early hours after Andrew made landfall, FEMA was waiting for direction from the president. The president was waiting for a written request for assistance from the governor of Florida. The governor was unaware that such a request was required.[98]

And then the finger-pointing started...

*"HURRICANE ANDREW; BREAKDOWN
SEEN IN U.S. STORM AID*

*WASHINGTON—Interviews with officials at numerous
Federal agencies suggest that there was a breakdown in
communication and coordination at the top levels of the
Government. The victims' anger at the delay burst into
the national news media on Thursday, creating a serious
political problem for [the] President. White House officials
did not realize the severity of the damage.... Even some
officials of the emergency management agency were
puzzled by the pace of the Federal response. Peg Maloy,
the agency's spokeswoman, said: "Something is wrong.
I don't know where things are breaking down. Nobody
knows where it's breaking down. I'd like to know myself."*[99]

—The New York Times, August 28, 1992

In the months and years leading up to Andrew, Governor
Chiles, or more accurately Governor Chiles's disaster team,
was under the grip of moral hazard. It was under the mistaken
notion that it needn't prepare too much for disasters because
it would be protected from their consequences. It hoped and
believed that FEMA would come and that it would bring the
cavalry with it. Unfortunately for the people of Homestead,
the cavalry didn't come.

the benefit of experience

As you might expect, Florida learned a lot from Andrew.
This experience, as well as that of the devastating storms
of the 2000s (including Jeanne, Dennis, Wilma, Ivan, and
Charley), relieved the state of Florida disaster team of any

illusions that it had regarding FEMA. FEMA's apparent success in Florida after Hurricane Irma in 2017 was due largely to the fact that Florida did not need a lot of help.

a fragile bromance

Many will argue that much has changed in the decades since that hot August night twenty-five years ago when Andrew slammed into South Florida. So, let's go back to that bromance between President Obama and his FEMA director, Craig Fugate. If, as Fugate said, "FEMA is not the nation's emergency management team—FEMA is only a part of the team," how is the team doing? Can we assume that, as *The New York Times* puts it, the "lessons have been learned"? Have we fixed the flaws the world has seen in past disasters?

The answer is, yes and no.

Again, it depends on which disasters you are talking about. While FEMA is much better at home-rule disasters, it has made little progress in its readiness for Maria-class disasters. The Maria-class disaster is a different kind of disaster, one that requires a completely different approach. The failed Hurricane Maria response shows clearly that FEMA, unlike the state of Florida, has not learned the lessons of Hurricane Andrew.

Early in the response to the next Maria-class disaster, these lessons will be revealed yet again. This is because most governors continue to believe that a national disaster system stands ready to assist them if a disaster gets too big for them to handle.

They believe that requesting a presidential major disaster declaration will bring FEMA, and when FEMA comes, it will bring the cavalry with it.

Here's the problem. Everybody believes that—except FEMA.

what exactly do you mean when you say "cavalry"?

For one clue, let's look more closely at FEMA's definition of "major disaster declaration:"

> *"A major disaster declaration provides a wide range of federal assistance programs for individuals and public infrastructure, including funds for both emergency and permanent work."*[100]
>
> —Stafford Disaster Relief and Emergency Assistance Act

Does that sound like the cavalry to you?

More than a hundred times a year, a governor appeals to the president for help with disaster relief. And so, every seventy-two hours, the help comes. But here's the thing: it doesn't come 'round the mountain on a white steed with the flag flying and the cavalry close behind. Instead of a national disaster response, FEMA just brings itself. FEMA's cavalry is a bunch of old, paunchy guys in polo shirts clutching checkbooks. Bringing up the rear is the new guy who was given the job of dragging in the threadbare carpetbag filled with rusty dental instruments.

These rusty dental instruments are the "federal assistance programs for individuals and public infrastructure" referred to by the President in his declaration (more about these later).

To be fair, FEMA does have the ability to move massive quantities of bottled water and packaged food. And it can bring other resources too, mostly through mission assignments to other federal agencies. Examples of this include emergency power generators and water pumps from the US Army Corps of Engineers and temporary field hospitals and medical teams from the US Department of Health and Human Services.

But, in the aftermath of a catastrophe, any help that FEMA brings will be too little and too late. As Chuck Hagan, the State of Florida's Unified State Logistics Chief and disaster logistics legend, once said to me, "We know that we are going to be on our own. For us, FEMA is just another vendor."

the myth of the system

Conventional wisdom holds that the United States of America has a national disaster system that can mobilize an immediate and massive response in the aftermath of a catastrophe. It turns out that, as with a lot of conventional wisdom, this is a myth.

Instead, we face a dangerous future with fifty-plus state disaster systems duct-taped together, with fifty-plus different structures, capabilities, and methods.

Individual states have responsibility for their own disasters and no responsibility for disasters anywhere else. And they are on their own when it comes to asking for help from other states. The problem is that each state is 'an empire within an empire' with relationships between them shifting with the political winds.[101] Although some states do reach across to help their neighbors during disasters, the process is ad hoc, haphazard, and slow. So, during the next Maria-class disaster, as with all past disasters, it will be the responsibility of the state to activate its own national disaster system.

FEMA has never taken on what is arguably its most important role: convening Governors. Rather than lead a process to bind the fifty-plus states together into a national team, FEMA prefers to work with each of them separately. Worse yet, FEMA takes on much of what the states should be doing while leaving to them the one thing that it should be doing itself.

there ain't no such thing as free money

We talked about how the promise of FEMA free money causes states to underprepare for disasters. Another, even more destructive effect of FEMA free money is the process itself. It is a massive administrative burden, both for the state and for FEMA, leaving no time for the real work that must be done.

FEMA's biggest free money programs are Individual Assistance and Public Assistance. The Individual Assistance Program, as its name suggests, is for individuals—whether they are homeowners or renters. It gives them money to fix disaster-related damage to their homes. The Public Assistance Program gives money to state and local governments to repair roads, bridges, public buildings, and parks. It also pays for the overtime worked by first responders and emergency managers during disasters.

Experts have long debated the wisdom of this approach. They question why the rancher in Wyoming should be forced to pay the salary of an emergency manager in Key West during a hurricane response. Economists argue effective risk management would call for the state that benefits year-round from a tropical climate to bear the cost of its hazards from tropical storms.

Beyond the dysfunctional incentives caused by Individual Assistance and Public Assistance, is the damage they do to our national preparedness. The IA and PA process (including client contacts, negotiations, paperwork, and field inspections) for the hundreds of active DRs is a massive task that takes up the lion's share of FEMA's time, leaving little time for its primary mission, preparing the nation for catastrophes.

Having abandoned its primary mission, FEMA leaves it to the states. Through the Emergency Management Performance Grant (EMPG) and other programs, FEMA gives yet more free money to states and locals to prepare for catastrophes. And, while EMPG is good free money, as we shall see (in the *All Disasters Are Local* Index described in Chapter 10), it is not nearly enough.

Every day all around the country, FEMA wraps its time and talent around a meaningless paperwork exercise that leaves it—and the nation—unprepared[102]. Thus, we have succeeded in nationalizing the local and localizing the national. FEMA is doing what states and locals should be doing and leaving to them the one thing that it should do itself. Somehow, we have managed to build the system exactly backwards.

Let's go back to Karl Metairie and his team in the New Orleans EOC. In those crucial early hours, when the nation needed to get big enough fast enough to manage the surge in human needs that the crisis brought, there were several response organizations all but disconnected from each other, working at cross purposes, and losing time.

We see this in the aftermath of every disaster with FEMA, the State, and the local government setting up in separate locations remote from the disaster. FEMA activates its Joint Field Office, the state activates a separate operations center, and the local government activates its Emergency Operations Center. So instead of one disaster response, we get three. This complex environment, with multiple overlapping jurisdictions and different missions, is massively dysfunctional.

In his prophetic book *Lights Out*, Ted Koppel makes clear that the federal government is not ready for the aftermath of an attack on the power grid. If a highly populated area was without electricity for a period of months or even weeks, there is no master plan for the civilian population. When he asked the Secretary of Homeland Security for the plan, the Secretary could only suggest that people keep a battery-powered radio. Koppel concludes that government is not prepared for such a catastrophe. The premise of this book is that our government is not prepared for a catastrophe of any type or cause. [103]

shouldn't we all be better at this?

"This is a wake-up call. People cannot depend solely on the Federal Emergency Management Agency. I think that we all have to sit down and figure out how to collectively improve."

—FEMA Administrator Brock Long on
Face the Nation, Sept. 3, 2017[104]

Of course, they don't just happen here.

Wherever it occurs, the aftermath of every catastrophe includes a period of finger-pointing as the players scramble to avoid blame for its failures; failures that come in the form of suffering on the part of children and families.

Take the example of Super Typhoon Yolanda. Early in the morning of November 7, 2013, one of the strongest storms ever recorded made landfall in the central Philippines. Super Typhoon Yolanda crushed Tacloban City with 145 mph winds and twenty-foot waves.

In an interview on Philippine television, the Mayor of Tacloban City said, "There is no Tacloban City right now"; journalists on the ground described the devastation as, "off the scale; apocalyptic."

Nearly every family lost someone; they came in from outlying provinces looking for relatives, especially children, who may have been washed away. The entire first floor of the Tacloban City Convention Center, which was serving as an evacuation shelter, was submerged by storm surge. People in the building who had come for shelter were caught off-guard by the fast-rising waters and drowned.

The aftermath was even more devastating. Five, ten, even fifteen long days passed before relief arrived too little and too late. As we watched the devastating scenes of families and children crying out for food, water, and medical care, we asked ourselves, "Shouldn't help have come sooner? Shouldn't the Philippine government be more involved? Shouldn't we all be better at this?"

In a television interview on November 12, CNN's Christiane Amanpour asked Philippine President Benigno Aquino III why it was taking so long to get help to the people of Tacloban City. This is what he said:

> *"..our efforts rely on the local government units... our system says that the local government has to take care of the initial response...and unfortunately [they] were simply overwhelmed by the degree of this typhoon that has affected us...."*

President Benigno's words mirror what many national leaders, including our own, have said in the early days of the government-led disaster response. That is because the system in the United States (and in many other countries with federal systems [e.g., Canada, Australia and New Zealand]) mirrors that of the Philippines. Federal officials use the Stafford Act and Home Rule laws to draw bright lines around their responsibilities. They tell the President that the state and local governments 'own the job' and that they are leaning forward to 'support' them with anything they might need.

For decades, presidents have been fed this myth by disaster professionals only to discover—too late in the job—the painful reality. President H.W. Bush discovered it in 1992 after Hurricane Andrew, thirteen years later his son learned it after Hurricane Katrina. In 2017, Donald Trump learned it after Hurricane Maria.

an emergency management agency
that doesn't manage emergencies

At first, the Trump administration seemed to be doing all the right things to respond to the disaster in Puerto Rico.[105] As Hurricane Maria made landfall on Wednesday, September 20, there was a frenzy of activity. The President called local officials on the island, issued an emergency declaration and pledged that all federal resources would be directed to help. But for four days after that—as storm-ravaged Puerto Rico struggled for food and water amid the darkness of power outages—Trump and his top aides went dark. Air Force 1 took the President for a long weekend at his private golf club. Neither Trump nor any of his senior White House aides said a word about the growing crisis.

A similar situation unfolded in August 2005, when President Bush was oblivious to the crisis spawned by Hurricane Katrina, spending day after long day at his 1,600-acre Prairie Chapel Ranch in Crawford, Texas. His staff didn't want to burden him with detailed information about the situation in New Orleans. As the crisis grew, Bush's aides decided they had to inform the president about it in stark terms. One of his aides put together a video showing scenes of hurricane-ravaged communities and showed it to the president. At this point, Bush decided he should cut his vacation short and return. He flew back to Washington on August 31, after twenty-nine days at his ranch.

In all of these cases, the federal government was sitting back, waiting for its "all disasters are local" policy to kick in.

The mission of emergency management is to get big enough fast enough to find and fix the surge in consequences that the crisis brings. FEMA thinks it is somebody else's job to do this. That is why it has never developed an ability to do the things that other emergency managers do, like find and fix impacts to seniors, individuals with special needs, children and families in the early hours of the crisis. In this way, FEMA is unique and wholly different from every other emergency management agency in the US.

Thus, we are destined to watch the next Maria-Class disaster, another Tacloban scenario, unfold in this country just as we did in Homestead in 1992, New Orleans in 2005, and Puerto Rico in 2017.

Then, instead of outcomes, we will see only finger-pointing. During the Hurricane Katrina crisis, the public and the media pointed at President Bush and FEMA Administrator Michael Brown. Brown pointed at Louisiana Governor Kathleen Blanco who, in turn, pointed at Mayor Ray Nagin and Karl Metairie in the New Orleans Emergency Operations Center. Ask them today how well that worked for them.

6 the first great machine

new york city shows us the way

> "A Sanitation Department salt spreader crashes
> through a wall and dangles three stories above the street,
> a Parks Department garbage truck falls into the East River,
> a crane falls onto Con Edison power lines, a boom truck
> falls onto a house, an earthquake shakes us, a hurricane
> strikes, and a tornado touches down in Queens. To some
> in our business this would be a lifetime of experiences.
> But for us this was just the last two weeks of August."[106]

—Deputy Commissioner John Scrivani's farewell
email to OEM, September 2011

the birth of oem

Born and raised in East Flatbush, Brooklyn, Rudolph William
Louis Giuliani defeated Democrat David Dinkins to become
the first Republican mayor of New York City in thirty years.
At the time, the city was reeling from a spike in crime and
unemployment from a recent nationwide recession. Giuliani

had promised New Yorkers that he would eradicate "the street tax paid to drunks and panhandlers...the squeegee men shaking down the motorist waiting at a light...the trash storms, the swirling mass of garbage left by peddlers and panhandlers, the open-air drug bazaars on unclean streets."[107]

After his inauguration in January 1994, Giuliani took the reins of an enormous government bureaucracy, bigger even than that of most US states.[108] Headquartered at City Hall in Lower Manhattan, New York City government was organized around a "strong mayor" system. Its more than three hundred thousand city workers were responsible for all city services and for enforcing all city laws. That centralized structure suited Giuliani perfectly, because as mayor, he was the boss of dozens of powerful city agencies, including fire, police, health and human services, housing, sanitation, buildings, water, and wastewater.

Giuliani brought a completely new approach to City Hall. He did not come up through the local political clubs and so did not owe a huge burden of favors. He was an outsider—a tough former federal prosecutor with an intuitive grasp of the laws of power and how to use them. First and foremost, he knew that without accountability, nothing got done. He quickly gained an encyclopedic knowledge of the machinery of his government.[109] He knew who was responsible for what, several layers down the New York City organizational chart. Giuliani and his lieutenants would reach deep into any agency at any time to get what they needed.

From the earliest days of his administration, Giuliani was frustrated with turf wars on the street between the New York City Police Department (NYPD) and the New York City Fire Department (FDNY). The battle between Big Blue and Big Red had begun in the 1980s and escalated throughout the decade and into the 1990s. To tackle the longstanding rivalry, he needed to project his authority into the field, but he couldn't always be everywhere. Giuliani needed an extension of himself, someone who could bring the boss to the street.

Two years into his first term, he heard about a small unit buried in the Police Department that oversaw something called emergency management. He plucked it out of the NYPD and brought it to City Hall. He renamed it the Mayor's Office of Emergency Management and hired a former IBM crisis manager[110] named Jerome Hauer to run it. Hauer reported directly to the mayor and quickly became an integral part of his inner circle of advisers. The Mayor's Office of Emergency Management, or OEM, began with a staff of just twelve. They were an elite group of seasoned veterans from the city's top agencies. One of the group's first tasks was to improve the coordination between FDNY and NYPD by specifying which agency had authority over what emergencies.

OEM came with a communications center called Watch Command. Watch Command's job was to look and listen, around the clock and around the world, for any sign of approaching danger. Watch Command got information from communications channels such as emergency radios, alert systems, breaking news, live video feeds from New York Harbor and the city's streets, and other watch centers around the state and the nation.

While Watch Command was Giuliani's eyes and ears on the world, OEM was his eyes and ears on the street. Hauer and his team, with an assortment of cellphones and pagers clipped to their belts, became fixtures at field emergencies, such as helicopter crashes, subway fires, building collapses, and water main breaks. At the same time, it worked its way into a leadership role in the city's response to major incidents, like the West Nile virus outbreak and Y2K.[111] When it showed up at a job, either on the street or at City Hall, OEM walked up like a boss.

OEM had what the old-timers called "juice." They knew the rules[112] and the players in the agencies who could turn the gears in the vast machinery of New York City government. They would call these players at all hours of the day or night with an "ask." An ask is a favor or an unusual request; it included things such as a front-end loader to move a fallen tree or a structural engineer to assess the integrity of a building wall, usually urgently needed in the middle of a cold winter's night. Everybody knew that an ask from Jerry Hauer was pretty much the same thing as an order from the mayor. The logic was simple: you could say no to Hauer, but Hauer was just going to call Joe Lhota[113], Guiliani's Deputy Mayor for Operations. The next call you would get would be from your boss after he or she had got a call from Lhota. Better to skip those last couple of steps and just do the ask.

Throughout his eight years in office, Giuliani pursued an aggressive agenda focused on crime control[114] and urban reconstruction.[115] His relentless focus on accountability worked.[116] He was credited not only with cleaning up city streets but with improving the overall quality of life in New York City.

In the fall of 2001, with Giuliani's second term in office coming to a close, the vast machinery of city government was running mostly on autopilot. When the sun rose on a crisp fall Tuesday in September, most of his senior lieutenants were working on their next career moves. But before that morning was over, everything would change.

9/11

OEM had a new director on September 11, 2001. Hauer had left the year before and had been replaced by Richie Sheirer. Sheirer had begun his career as an FDNY dispatcher and had worked his way up through the ranks. After he took over the top job at OEM, he grew it to more than seventy full-time staff.

OEM had a new headquarters, too. Two years earlier, OEM staff, along with a state-of-the-art EOC, had moved to the 23rd floor of World Trade Center 7, just across the street from the North Tower of the World Trade Center complex. At nearly 1,400 feet,[117] the North Tower was the tallest building in the world.

At 8:46 a.m. on September 11, five hijackers crashed American Airlines Flight 11 into the northern face of the North Tower. After burning for an hour and forty minutes, the North Tower collapsed. Massive pieces of steel and concrete rained down onto World Trade Center 7, piercing its exterior and igniting fires on at least ten floors.

By then most of the OEM staff was out of the building. They were on the street, supporting the response at the FDNY command post or moving equipment into place. All were caught in the avalanche of dust and debris from the collapsing towers; some barely escaped with their lives.

An hour later, as the building burned "like a giant torch,"[118] FDNY abandoned its last efforts to save it. Heat from the fire expanded the girders in its steel floor, causing the beams to buckle and pull away from the structural columns. The east penthouse began to crumble in the late afternoon. The remainder gave way, and World Trade Center 7 collapsed completely at 5:21 p.m., hurling its iron columns into the ground like red-hot spears.

Watch Command relocated to OEM's mobile command bus and kept working throughout that long day. But the loss of the EOC meant that OEM, and the city, was without its brain. Late in the afternoon, OEM created a temporary EOC at the Police Academy on East 20th Street in the Gramercy neighborhood of Manhattan. From there, a traumatized OEM spent the remainder of the day and that night pulling itself together.

Many in New York wondered whether the sun would rise again on Wednesday morning. As it finally did, OEM and its government, voluntary agency and private sector partners gathered in the makeshift EOC in the second-floor library of the Police Academy. The Port Authority reported that 15,000 people were in the towers when they collapsed. The conflicting images, of so many lost and yet so many potentially still alive, created an agonizing sense of urgency. There is nothing that anybody in the room would not have done to save even one life. They settled for doing the only thing that they knew how to do: They went to work.

OEM Deputy Director Henry Jackson went looking for a replacement site for his EOC, and quickly settled on the New York Passenger Ship Terminal on West 53rd Street. Built in 1935 for luxury ocean liners,[119] Pier 92 jutted out 1,100 feet into the Hudson River. Jackson was familiar with the site, because it had been scheduled to host OEM's bioterrorism exercise (dubbed Operation TriPOD) the next day after the attack, on September 12. Although not ideal, the space was big enough, easily accessible to the World Trade Center site, and available.

In one of the most extraordinary technology deployments in history, Jackson and his team converted the space into a fully operational EOC in less than seventy-two hours. On

September 14, the massive room, with over 75,000 square feet of space and twenty-six-foot-high ceilings, burst into life. Dozens of federal, city, state, and local agencies,[120] over 150 in all, came together there to confront the crisis, and New York's first "Great Machine" was born. As it surged to address all the issues New York City was facing, OEM was transformed from a tactical team focused on street-level emergencies to an elite team of professionals who could handle any type and size of crisis.

the first great machine

A laser focus from the nation—and the world—was on the response. "Today," the French newspaper *Le Monde* announced on September 12, "we are all Americans." The impact of 9/11 seemed to affect nearly everyone, everywhere, and it elicited an outpouring of solidarity and support.[121] Hundreds, even thousands, of people descended on New York to show that solidarity and to help in any way they could.

The EOC itself became a magnet for presidents and prime ministers, movie stars and sports celebrities, business leaders and technical experts. At the same time, the agency representatives struggled, mired in their emotional basements, with the surge of issues brought by the crisis.

As the agency representatives worked through a mountain of problems, they hadn't the slightest interest in politicians or movie stars, but many of those business leaders and technical experts had resources they needed, along with a passionate desire to help. In those early days—when fourteen hours would pass in a New York minute—OEM would sometimes bypass the city's procurement systems. For anything they needed, help was just an ask away. So, they asked.

They asked a special effects expert who made rain on movie sets to help keep the dust down, underground mining experts for advice about fighting fires in deep cavernous voids, and safety experts from the world's largest construction company to create a health and safety plan for rescue workers.

It wasn't only OEM that could make an ask; everyone in the Great Machine was empowered to do whatever it took to solve a problem or fill a need. On the morning of September 12, was assigned to the EOC as the New York City Health Department representative. Among our multitude of tasks was to find a better way for rescue workers to wash their hands. A couple of those fourteen-hour days passed before I finally managed an internet search. I found a company called PolyJohn Enterprises and called the number on its website. A few transfers later and I was talking to the company's owner, George Hiskes. I told him who I was and asked for his help.

"I need hand-washing stations," I said.

"Well, you got 'em. How many?" he asked.

PolyJohn Enterprises is located in the working-class town of Whiting, Indiana, on the southern shore of Lake Michigan, two miles from Chicago's South Side. It was four o'clock on a Friday afternoon. They were in the process of shutting the factory down, and everyone was looking forward to a beautiful fall weekend at the beach or the football field. Hiskes walked into the plant, gathered everyone together, and told them the story.

"Any questions?" he asked.

There was just one: "How many do they need?"

Instead of shutting down—instead of going home to their families and their football games and their picnics at the beach—the PolyJohn staff stayed. And they kept the plant going all weekend. On Monday morning, three tractor trailers with Indiana plates arrived in Lower Manhattan. They worked throughout the day, dropping a hundred portable hand-washing stations equipped with soap, paper towel dispensers, and thirty-gallon water tanks around the perimeter of the sixteen acre site that became known as Ground Zero. Hiskes never asked me for so much as a credit-card number.

Thus, the power of making the ask.

Over the next ten months, dozens of people and companies from around the country and the world pitched in to help. OEM saved every contact, every phone number and email address, in its rolodex.

the eoc gets really busy

The New York City EOC was the epicenter of the crisis. In those early days, the cavernous warehouse was so busy and so loud that you could hardly hear yourself think. And the busiest room on the planet was about to get a lot busier.

On the Tuesday after the attacks, five letters were dropped into a mailbox at Ten Nassau Street in Princeton, New Jersey, just across the street from the Princeton University campus. The envelopes were addressed to ABC News, CBS News, NBC News, and the *New York Post*, all located in New York City, and to the *National Enquirer* in Boca Raton, Florida. They contained a coarse brown granular material[122] that turned out to be a particularly lethal form of weaponized anthrax spores. Soon five people would be dead.

A few weeks later, on November 12, an Airbus A300 en route to Santo Domingo in the Dominican Republic crashed shortly after takeoff from John F. Kennedy International Airport in Queens. A rudder failure caused the plane to pitch downward into the Belle Harbor neighborhood in that same borough of Queens. All 260 people aboard the plane and five bystanders on the ground were killed. It was the second-deadliest aviation accident ever on US soil.

As 2001 ended, the world seemed to be coming unglued and OEM was sagging under the burden of a multitude of crises. It was getting crushed, under siege by a herd of dragons to be slayed and wave after wave of white-hot problems to be solved.

But the OEM staff held together as a team. They called everybody in the OEM rolodex who could help them. Few said no. They kept a calm composure. They told everybody what was happening; they set the battle rhythm. They answered the questions that nobody else would answer and solved the problems that no one else would solve.

Rather than break, they stood their ground—on a thousand-square-foot podium—smack in the middle of the chaos. And they got through it, working amidst the deafening tumult, twenty-four hours a day for nearly a year. 2001 was the year of the worst tragedy in the history of New York. It was also the year that OEM—and the city of New York—triumphed.

new boss

Mark Green, a former public interest lawyer and the city's Public Advocate, was heavily favored in the Mayoral primary election that was scheduled for the morning of September 11. The primary was postponed, and by the time it got back on track, the political landscape—like everything else in New York—had changed. Everyone knew that Michael Bloomberg didn't stand a chance in a city where Democrats outnumbered Republicans ten to one. But, with Giuliani's endorsement and sixty-nine million dollars of his own money, he defeated Green fifty percent to forty-eight percent in the closest mayoral election in a century.

OEM was getting a new boss with a whole new worldview. Bloomberg was aggressive like Giuliani, but he was not heavy-handed and was virtually apolitical. He, too, demanded results, but the team he surrounded himself with used data instead of street-smarts and intuition to understand issues and solve problems.

and the crises kept on coming

I left the Health Department in 2003 but returned to city government in February 2006 as Deputy Commissioner at OEM. During my first week on the job, we activated the EOC as the biggest winter storm in New York City history— dubbed the Blizzard of '06—buried the Northeast under blowing, drifting, thigh-high snows that crippled transportation, knocked out power, and disrupted life for millions in fourteen states.[123]

Before the year was out, we would face another anthrax scare, an Upper East Side townhouse explosion and collapse, a ten-day power blackout during a summer heat wave, and an airplane crashing through the 30th-floor windows of a condominium tower on East 72nd Street.

The following year, a steam pipe explosion killed one and wounded 20 others near the corner of Lexington Avenue and East 41st Street in Manhattan.

The year after that, tower cranes began falling all over the east side of Manhattan and US Airways Flight 1549 (the "Miracle on the Hudson") landed in the water off West 42nd Street after both engines failed.

In April 2010, swine flu was detected in students at St. Francis Preparatory School in Queens.

In September of that same year, a tornado hit Brooklyn, Queens, and Staten Island, and in August 2011, Hurricane Irene made landfall on Coney Island.

And, in our spare time, we managed blizzards, heat waves, and major building fires.

In my nearly eight years there, we responded to nearly five hundred incidents every year, with large-scale incidents occurring on an almost monthly basis.

insight that can be gained only through experience

The post-9/11 period in New York was unique in the history of the modern city, when we learned the nature of catastrophes and how to prepare for them.

We activated the Great Machine, centered in the New York City EOC, more than ten times a year, every year. Our time inside the parallel universe showed us what it takes to work there and how hard it is to get things done. We learned that success or failure is determined by what we do and don't do in the first hours of the crisis.[124] We learned the hard way that excuses are worthless—and that because there can be no excuse for failure, failure cannot be an option.

Most important, we came to know the "spectrous fiend" that is the crisis and the weaponry that it has, especially its most fearsome weapon, *surge*. We already knew how to do a lot of things at the same time—but surge forced us to figure out how to do everything all at once.

We were lucky in so many ways. Since 9/11 we doubled in size, from 70 staff to 150. We had enough staff to field three[125] duty teams. Even though it seemed like we were always exhausted, we were not incapacitated. And, with each crisis, we built our network and our tools.

We became experts at getting big fast, by pulling people and teams away from their day jobs in other governments, the private sector, and nonprofit and faith-based organizations. We brought them together with a unity of purpose into our Great Machine (also known by its formal name, the *incident organization*) to collaborate, innovate, solve problems, make decisions, and act in the moment.

What we didn't know was that, at the same time and ten thousand miles away, another team with a vital mission was learning the same lessons....

the other parallel universe

On March 20, 2003, a "shock and awe" bombing campaign signaled the start of the invasion of the US-led coalition into Iraq. Three weeks later, after twenty-one days of heavy fighting, 180,000 troops from the United States, the United Kingdom, Australia, and Poland overwhelmed Iraqi forces and brought down the government of Saddam Hussein.

On May 1, 2003, President George W. Bush addressed the nation from the deck of the aircraft carrier USS *Abraham Lincoln* in front of a giant sign that read, "Mission Accomplished."

Unfortunately for the president, and the nation, the majority of casualties from the Iraq war, both military and civilian, were yet to come.

Soon after the invasion, Al Qaeda in Iraq (or AQI) began a campaign of suicide-bomber attacks targeting UN representatives, security forces, and Iraqi civilians. As the weeks passed, the insurgency continued to expand, injecting chaos into the parallel universe of the war zone.

The objective of AQI's ruthless leader, Abu Musab al-Zarqawi, was to deepen the sectarian conflict that was the heart of the Iraq War.

into the chaos

In September 2003, Major General Stanley McChrystal was assigned command of the Joint Special Operations Task Force (hereafter, the "Task Force") that was headquartered at the Joint Operations Base in Balad, Iraq. General McChrystal was given the unenviable job of finding out who was behind the rising tide of violence and putting an end to it. But by early 2004, McChrystal realized that despite its huge advantage in numbers, equipment, and training, the Task Force was losing the fight.

Overcoming this involved more than increasing the number of precision Special Forces raids. It required solving a multitude of complex challenges, not the least of which was between the ears of the Task Force members themselves, in the form of the conventional wisdom that had been built up in the US military over the previous hundred years. McChrystal needed to reject that conventional wisdom and find another way to confront this new threat.

The enemy was engaged in modern-day guerrilla warfare. It was made up of small independent teams, attacking quickly and then disappearing into the civilian population. These teams displayed many of the good traits that define small teams: they were agile and adaptable; they shared a common purpose and a common situational awareness. And they were empowered to act.[126] McChrystal knew that to prevail, the Task Force would have to mirror these traits and then use them on the streets of Ramadi and Mosul.

He began to completely remake the Task Force in the midst of the war, to scale the adaptability of the small team up to the enterprise level. To do this he needed to transform it from a team of soldiers and officers into a team of teams.

This new organization would be founded on the concept of transparency, so that everybody would know what was going on all the time. This common situational awareness enabled teams that previously existed in separate silos to be brought together around the same mission, to work collaboratively, to innovate. Decision-making authority was pushed down to the team level, allowing the members to act quickly.

> *"Individuals and teams closest to the problem armed with unprecedented levels of insights from across the network offered the best ability to decide and act decisively."*[127]
>
> —General Stanley McChrystal, *Team of Teams*

Harnessing the power of teams allowed the Task Force to adapt quickly to the rapidly changing environment inside the parallel universe. On the streets of Najaf and Baghdad, the teams worked together to invent solutions and act in the moment, rather than waiting for orders from the top.

The Task Force began to work a process that centered on the daily operations and intelligence briefing. The O&I Briefing was an "all-hands" video conference that started with a detailed report of the current conditions on the battlefield, followed by a summary of the mission objectives and an open discussion of obstacles and unmet needs. If this sounds strangely familiar, that is because it is the format followed during many typical business meetings. But it was much more than that. Like the Duty Team calls at OEM, the O&I Briefing was the glue that held the Task Force together. It set the battle rhythm, pumping out information about the entire scope of the operation to all members of the Task Force and partner agencies and, at the same time, allowing them to contribute their observations and ideas.

the great machine is a team of teams

Like *The Black Swan* before it, *Team of Teams* has become required reading in emergency management agencies across the nation. Disaster professionals in New York City were struck by the similarities between the parallel universe of the disaster zone and the parallel universe of the war zone. In New York City, the crisis was our AQI, the enemy that sought to destroy us. And, like General McChrystal, we found that the biggest and best weapon in our arsenal was the Great Machine.

Like the Joint Special Operations Task Force in Iraq, our Great Machine is a team of teams. It is composed of people and resources coming together to work a process—fast, flat, and flexible,[128] combining transparent communication with decentralized decision-making.

Some disasters require dozens of teams focused on different aspects of the disaster—search and rescue, damage assessment, evacuation, sheltering, logistics, debris removal, disaster assistance, fatality management, feeding, and on and on.

Even though every team is empowered to feed and care for itself, for the days and weeks of a big response, the job of the Great Machine is to get them whatever they can't get for themselves. If a team needs leadership, it assigns it. It solves the problems they can't solve and moves the obstacles they can't overcome. It gets them the stuff—from industry experts to specialized vehicles or equipment—they need to do their job. If there is information or orders or approvals they can't get, it will get them. For new problems for which there is no plan, it gets more people in and creates new teams.

The Great Machine creates trust—trust in the plan and confidence that we will not fail. It doesn't wait; it anticipates. It creates a collective dynamic that empowers teams to run at, not away from, problems. The OEM Duty Team calls, as with the O&I Briefings in Iraq, force them to think: "What is happening? What are we doing? What do we intend to do? What can we do now to get ahead of the curve?"

Finally, the Great Machine tells everybody what is going on: field teams to agency headquarters to city hall to the children and families trapped within the parallel universe. It tells them what life is like within the parallel universe, what we are doing about it, what we are not yet doing, and why.

People think that government has some innate ability to respond to disasters. Nothing could be further from the truth. Governments are slow-moving creatures of habit, ill-suited to the demands of the parallel universe. The Great Machine is the secret sauce, an instant bureaucracy that supercharges the government-led response.

the myth of crisis leadership

Another insight OEM had at roughly the same time as the Task Force involves conventional wisdom about leadership in crisis.

In Iraq, McChrystal knew that the complexity and scale of modern war fighting would always exceed the ability of any one person to comprehend and direct. His new approach required changing the traditional conception of the crisis leader. Rather than exerting so-called command and control (also known as micromanagement), McChrystal's leaders had the job of rallying the troops and enabling their clear thinking and execution.

> *"The temptation to lead as a Chessmaster controlling each move of the organization must give way to an approach as a gardener enabling rather than directing."*[129]

During a disaster, the Great Machine communicates a list of clear objectives for every operational period. At OEM we called those objectives Commander's Intent. Everyone, at every level of the organization, is empowered to say yes to everything as long as it falls within the boundaries of Commander's Intent. The message is, "Do what is right, not what you have a right to do."[130]

coastal storm plan

In the years immediately after 9/11, OEM kept a strong focus on terrorism. But by the time Hurricane Katrina struck near the Mississippi–Louisiana border early in the morning on Monday, August 29, 2005, our focus had changed.

Katrina was a catastrophe, a Maria-class disaster, with impacts that exceeded the combined ability of state[131] and local governments to respond. It was a game-changer that spanned borders, affecting multiple states simultaneously and demanding resources far beyond what was immediately available. The surge in humanitarian needs across a vast affected area cried out for a massive response. But with no Great Machine and nobody in charge, thousands of individuals and families were left to fend for themselves.

Some are baffled by the widely-held belief that the Hurricane Katrina response was "a colossal failure of government at every level—federal, state, and local."[132] They insist that thousands of people worked hard during that response, running massive operations on the ground. While this is true, the failures of Katrina stem from the fact that these things didn't happen fast enough. As with Andrew before it and Maria after, the government response got a couple of days behind the job and never caught up. When help finally did arrive, a lack of communication and coordination resulted in chaos, with voluntary groups working at cross-purposes with first responders; local, state, and federal agencies; and the National Guard. Some people got too much, while others got nothing.

At OEM, we became students of Katrina, learning the lessons of its failure. We read every after-action report[133] and used them to create a blueprint for our Coastal Storm Plan.

The massive New York City Coastal Storm Plan was built to immediately activate and simultaneously orchestrate a wide range of field operations, such as evacuation, sheltering, logistics, debris management, post-storm commodity distribution, and healthcare facility evacuation. To build it, we convened the Great Machine. We brought everyone together into the parallel universe to face the worst-case scenario. Then we practiced it, using daylong simulations of devastating hurricanes.

We leveraged the power of making the ask, using every contact in our network. We welded problems to teams which prompted them to build more teams on their side. We drilled and planned and drilled some more. We peeled the onion and asked ourselves what we would do. We worked through all of the impacts, including storm surge, stranded people, wind, evacuating hospitals, rain, downed trees, trapped cars, power failure—you name it.

After we lived the scenario together, we learned lessons and used them to refine our plan. So, when Hurricane Sandy came, we were ready. And, as luck would have it, the Sandy track turned out to be identical to the worst-case scenario track on which we had based the plan.

superstorm

The largest Atlantic hurricane on record (with winds spanning nearly 1,100 miles), Sandy moved ashore near Brigantine, New Jersey, late in the evening of October 29, 2012.

Sandy's arrival coincided with an astronomical high tide,[134] and brought an unprecedented storm surge all across the Atlantic-facing shores of New York City.

> "The storm's angle of approach put New York City in the path of the storm's onshore winds, the worst possible place to be. These winds pushed the storm's massive surge—and its large, battering waves—directly at New York with punishing force."[135]
>
> —New York City Special Initiative for Rebuilding and Resiliency, "Sandy and Its Impacts," June 2013

Hurricane Sandy brought with it a chaotic mix of death, destruction, and disruption. Its wall of water[136] combined with hurricane-force winds to knock out power across much of the city, leaving nearly eight hundred thousand people in the dark.

The onslaught of water rose up over beaches, boardwalks, and bulkheads, flooding the subway system and most of the roadway tunnels in and around Manhattan. Thousands of homes and businesses were flooded. The floodwaters contacted live electrical lines and sparked a fire in Breezy Point, Queens, that destroyed over a hundred homes.

After the backup power system failed at NYU Langone Medical Center, nearly a thousand staff and first responders carefully and methodically lowered over three hundred patients through the stairways of Tisch Hospital to ambulances waiting along the dark streets. Other hospitals, including the city's Bellevue Hospital Center, were also closed and evacuated.

new york's team of teams

The week before landfall, OEM had turned on the Great Machine and gathered the team of teams. By the time Sandy arrived, it was already leading a multitude of large-scale field operations, all at the same time.

The Logistics Team—led by OEM—was reaching deep into "the system" to find what we needed and get the right stuff in the right place at the right time. In the first week, it deployed 230 generators to hospitals, nursing homes, and public housing developments.

The Dewatering Team—led by the Army Corps of Engineers, the US Navy and the MTA—was getting giant pumps through debris-filled streets to pump out flooded roadways and tunnels and get the subways running.

The Debris-Removal Task Force—led by the New York City Department of Sanitation—was collecting millions of tons of debris.

The Downed-Tree Task Force—led by the New York City Parks Department—was removing more than twenty thousand trees damaged by the storm.

The Emergency Sheltering Team—led by the Department of Homeless Services—flipped the switch and, in less than twenty-four hours, transformed more than a hundred public schools into emergency shelters. The people staffing these makeshift shelters were city employees and volunteers who had been trained for this job. At its peak, the shelter operation housed more than eight thousand people.

sandy's most fearsome weapon

During those long days in late 2012, the epicenter of the crisis was OEM Headquarters at Cadman Plaza in downtown Brooklyn. The brainchild of Deputy Commissioner Henry Jackson, the 65,000-square-foot facility boasts a media briefing room and a state-of-the-art Watch Command and Emergency Operations Center.

It was immediately filled to overflowing with every office, hallway, and closet, from the basement up, crammed with teams. We brought in temporary trailers and parked them outside. We put more teams in other places all around the city, at MetroTech in Brooklyn, at Citi Field, and at the OEM warehouse in Queens. We missed the vast real estate of Pier 92.

Because we were not getting get big enough fast enough.

Our focus in those early days was the response, all the complex operations we needed to execute on a vast scale in the days just before and just after the hurricane's landfall. Although our recovery plans were operational and robust, the OEM staffers who had led those teams were stuck in response mode—dewatering, for instance, and emergency power and sheltering. In the midst of the surge, we ran out of leaders for all of our recovery teams. So, despite nearly six years of preparation, we were getting behind the job. We had the teams and the plans, but we were short on team leaders.

Fortunately, there were plenty of team leaders left in New York City government. And lots of them started to arrive, first and foremost from City Hall.

Deputy Mayors Cas Holloway and Linda Gibb rushed in with their teams, along with dozens of talented staff and experts from inside and outside the city.

Holloway had been with us every step of the way. As our primary contact at City Hall, he was in the Situation Room from the first weather forecast. He knew our operations almost as well as we did. An early champion of the Coastal Storm Plan, he got the funding and strong-armed commissioners into committing more resources (than they wanted to) to the cause.

In the run-up to the storm, Holloway and Gibb began to pull in the best managers and brightest minds in New York and from around the world.

They supercharged our information management processes. And we built teams around these new managers, especially for key recovery operations, such as food distribution, neighborhood recovery centers, and family case management. One of Gibb's teams turned the Situation Room into a command center for a food and water distribution operation that gave out more than 2.1 million meals ready-to-eat (or MREs) and more than 925,000 bottles of water in the affected areas. Another team delivered food, water, and other goods directly to residents in their homes. A third team focused on disaster case management for affected individuals and families who needed immediate help.

aftermath

By any measure, Sandy was unprecedented, affecting so many lives. At least fifty-three people died in New York because of the storm. Thousands of homes and an estimated 250,000 vehicles were destroyed, with economic losses in New York City estimated at over 19 billion dollars.

And then, as the water receded, New Yorkers dealt with the aftermath:

> *"It didn't take much time before everything turned darker as the real misery became more apparent.*
>
> *Fury mounted with every hour that electricity and heat and food failed to arrive.*
>
> *In Alphabet City, in Red Hook, out in Staten Island, there were people who needed to fill buckets from hydrants or scrounge from dumpsters.*
>
> *The news from the outer-boroughs was especially grim; people were fighting over gasoline..."*[137]
>
> —*New York Magazine*, November 4, 2012

Even as I write this, individuals and families, and in some ways the city itself, are still recovering from this devastating natural disaster and will continue to do so for years.

the anti-katrina

The Hurricane Sandy response was by no means flawless. Public housing, especially in the Rockaways, Coney Island, and Red Hook, were especially hard hit. The storm surge picked up the beach and moved it into the boiler rooms (sand dunes and all), impacting 80,000 public housing residents in more than four hundred buildings. It took too long to solve that complex challenge and get the heat and power back on in those buildings. Hospitals and nursing homes were caught without power and forced to evacuate amid the storm. A gasoline shortage dragged on and on, and not enough was said about it. We had more glitches and snafus from the very first minutes in the shelter system. And, of course, a lot has already been said about the housing recovery.[138]

Despite its flaws, the New York City Coastal Storm Plan, built in the aftermath of Katrina and using its lessons as a blueprint, allowed us to avoid the worst mistakes of the failed Katrina response seven years earlier.

And, as it did in September 2001, OEM turned on the Great Machine and stood its ground—on a tiny oval podium in the EOC—smack in the middle of the chaos. It worked in that parallel universe twenty-four hours a day for nearly twenty weeks.

We told everybody what was happening; we set the battle rhythm.

We answered the questions that nobody else could answer and solved the problems that no one else could solve. And, because of the heroic effort of thousands of first responders, healthcare workers, mass transit workers, power field teams, and others, we did not fail.

When the peak of the job passed, the mood in the EOC began to improve. As we started to get our arms around the most difficult problems, and the intensity edged down ever so slightly, there was a sense of relief. The OEM team had worked hard, harder than they had ever worked in their lives. And, at the end of the day, they knew that they had done a good job. In the run-up to the storm, most of us were wondering why we had ever decided to get into the disaster business.

But, as the long response phase wound down and we began to see the light at the end of the surge, some of those same people started to think that this was the best job in the whole world.

part III | *fault lines*

7 going native

bad things happen when you don't plug into the great machine

"THROUGH THE BLOODY September
twilight, aftermath of [a hundred] rainless
days, it had gone like a fire in dry
grass..."[139]

—William Faulkner, "Dry September"

***

it started on fish ranch road

The summer of 1970 was the hottest anyone could remember. In the foothills and mountains away from the coast it hit a hundred degrees, day after day. California has seen its share of legendary droughts, but as the summer wore on, it kept not raining. Not a drop of rain since June.

Then came the Diablos—those high, hot winds that blow west from the inland deserts, sometimes gusting to hurricane strength. They blew all day long, transforming the forests and everything in them—shrubs, grass, bark, branches, even the soil itself—into fuel.

September rolled around. The wind kept blowing, and the terrain kept getting drier. When the humidity dropped below two percent, firefighters sensed the epic battle ahead. They tied up loose ends, poured coffee, and settled in, waiting for the other shoe to drop.

And then, finally, it did...

> *"OAKLAND, Calif. Sept. 22, 1970. A wind-driven fire, believed to be set by arsonists, roared out of the Berkeley Hills today..."*
>
> *—The New York Times*

Someone set a match to tinder-dry grass along Fish Ranch Road in the Oakland Hills east of the University of California, Berkeley campus. Within minutes, the flames, feeding on dry coyote brush and pine trees and whipped by those Diablo winds, swept to the ridgetop and leaped into homes perched on the steep hillside above San Francisco Bay.

In less than two hours, thirty-six homes were gone, and thirty-seven others were aflame. The heat was so intense that the houses exploded before the fire even got to them. And it was just getting started. It would be two weeks before it was done. By then, 580,000 acres would be burned, sixteen people would be dead, and 722 homes would be destroyed.[140]

After thirteen days of fire, firefighters managed to stabilize the disaster. It ended slowly, stubbornly, when the fire boss of the 34,000-acre Meyers Fire in the hills north of San

Bernardino announced that his fire had been wholly surrounded by a cleared line.

The firefighters did a remarkable job considering the conditions they faced. Despite their Herculean efforts, the public was stunned by the scale of the devastation. The fire services were criticized for what was widely perceived to be a mismanaged response.

Formal investigations into the disaster found that a series of mistakes had compounded the disaster. These were not tactical mistakes or a lack of resources but management and communication failures.

To find a way forward, President Nixon created the National Commission on Fire Prevention and Control. In May 1973 it issued a report entitled "America Burning," which concluded that "fire is a major national problem."

Congress ordered the US Forest Service to fund a five-year research program called the Firefighting Resources of Southern California Organized for Potential Emergencies program, or FIRESCOPE.[141] The FIRESCOPE research team concluded that "because the fire disaster was reaching such widespread proportions and involved the firefighting apparatus of so many separate fire departments, there was need for a coordinating body."

So, FIRESCOPE created a new fire management system. It called this new system the Incident Command System, or ICS.

putting ics to the test

Within ten years, ICS would be in use throughout Southern California and spreading across the country.

Because it worked.

It was simple to learn and use and versatile enough to bring order to the chaos of any type of disaster, not just wild-fires. And it was scalable, meaning it could handle everything from a car accident to a catastrophic earthquake.

In New York City, OEM was an early adopter of ICS because it knew that in the parallel universe, ICS was the key to managing the massive city bureaucracy and especially its longstanding rivals, NYPD and FDNY.

In the fall of 2001, OEM was preparing to put ICS to the test in a citywide disaster exercise.

As in most other big cities and all states, New York City had a plan for how it would distribute prophylactic (or pro-tective) antibiotics or vaccines to people during epidemics or after a bioterror attack. Such medications could have to be provided to every one of the 8.4 million residents of the city, and quickly.

The citywide exercise was dubbed Operation Tripod, with "Tripod" being a sort of acronym for "trial point of dispens-ing." According to the exercise scenario, terrorists had released a large quantity of an anthrax "agent" in the subways. Pier 92, at the New York Passenger Ship Terminal on West 53rd Street, had been outfitted as a massive point-of-dispensing, or POD, site. OEM had recruited hundreds of Police Academy cadets and Fire Department trainees to act as civilians who had come to the POD for their medications. It had even bought 70,000 M&Ms to be handed out as fake medicine.[142]

Operation Tripod was scheduled for Wednesday, September 12. High-profile VIPs from around the city, including Mayor Giuliani, the police and fire commissioners, the FBI, and FEMA, were invited as observers.

On Tuesday, September 11, many of the OEM staff had come to work early to prepare for the exercise. They were expecting a busy day.[143] Nobody had the slightest inkling of just how busy it was going to get.

attack

It began at 8:46 a.m., when five hijackers crashed American Airlines Flight 11 into the northern face of the World Trade Center's North Tower. It ended when World Trade Center 7 collapsed at 5:21 p.m. and the Twin Towers, along with more than a dozen other nearby buildings, lay in ruins.

In the early hours after the attacks, as firefighters, construction workers, and volunteers dug through the smoldering pile of debris, an alphabet soup of federal, state, and city agencies poured into Lower Manhattan. Many of these agencies—including EPA, OSHA, HHS, NIOSH, PESH and COSH (Environmental Protection Agency, Occupational Health and Safety Administration, Department of Health and Human Services, CDC's National Institute of Occupational Safety and Health, New York State Department of Labor's Public Employees Safety and Health Bureau, NYC Citywide Occupational Safety and Health)—were responsible, in one way or another, for the health and safety of people affected by disasters.

But in those dark days, the stakes—and the risks—were sky high. The media placed a laser-focus on Ground Zero. Every act, every statement, was magnified and overanalyzed. Mistakes could be career-ending.

In the case of the workers on the pile and the residents of the neighborhood, the agencies wanted desperately to be involved, but they were equally desperate to avoid any mistakes for which they could be blamed.

Issues like worker safety and environmental health were burning white hot in Lower Manhattan in the early days after the attacks, so hot that the bureaucrats who, on a normal day, were responsible for these things suddenly wouldn't touch them.

the tail of the response

One of the reasons you don't see a lot of emergency managers in disaster movies is because, instead of "running toward the boom," they run into carpeted conference rooms to huddle over laptops with cellphones stuck to their ears.

The most important work, by far, in the parallel universe is done by police officers and firefighters and humanitarian aid workers in the field. This is known as the tactical level, or "teeth," of the response.

At the same time, a team of teams is gathered in and around operations centers, supporting the teeth of the response with critical information and resources and solving the problems that the tactical-level can't solve on the ground.

From the outside, the EOC—the "tail" of the response—looks boring. But that tail work can have a profound effect on outcomes in the field.

the morning after

In the early morning after the attacks, as the Associate Commissioner for Environmental Health at the New York City Department of Health, I was assigned to the makeshift Emergency Operations Center in the second-floor library of the city's Police Academy on East 20th Street, six blocks above the evacuated city—the so-called "Red Zone."

Some of my colleagues from the Health Department had traveled downtown overnight and were still at the scene. Their reports narrowed the Health Department focus that morning to the hazards faced by the first responders fighting fires and digging in the rubble in and around the massive site.

We knew that standard protocols called for every workplace to be assessed for hazards to the workers before any job could start. According to those protocols, once the hazards are known, workers are given the equipment and the training needed to keep them safe. That is *standard practice*. Unfortunately, in the case of Lower Manhattan on September 12, 2001, standard practice was out of the question.

Overnight my boss, Dr. Ben Mojica, Deputy Commissioner of Health, had issued a safety advisory recommending that workers at Ground Zero be outfitted with N95 respirators (close-fitting face masks that cover the mouth and nose), hard hats, neoprene gloves, goggles, and steel-toed boots. When Mojica read the advisory to me over the phone, I was puzzled:

"Ben, why did we issue this? FDNY has its own Safety Chiefs. They don't need us to tell firefighters to wear steel-toed boots."

He explained that all kinds of people were downtown trying to help, including construction workers who had been working nearby and civilians who showed up out of nowhere.

"Who's controlling all of those people? Fire?"

"FDNY is in charge," Mojica said. "But it's pretty chaotic right now."

I hung up and turned to the flood of issues that were landing on my desk. CDC had dropped a Push Package[144] of drugs and medical supplies at Kennedy airport and we worked to get it downtown to the World Trade Center site; we got pulled into a conversation about rescue workers being bitten by rats; we were besieged by requests from people trying to get downtown.

Just before 8 a.m., Randy Price, vice president of environmental health and safety at Consolidated Edison (the power company), pushed his way over to my desk. He handed me a piece of paper with four lines of data on it. The top line was a laboratory result from an air sample collected the day before at Ground Zero. A Con Edison worker wearing a sampler had walked the perimeter of the site, just a couple of hours after the buildings fell. The other three lines were lab results for samples of debris from the street that the worker had collected at the same time. The data showed low levels of asbestos in the air and the debris.

Laboratory results that are above legal thresholds are easy to interpret. Those that are not are harder. Since these results fell into that second category,[145] I wanted Con Edison's interpretation.

"Okay, we got less than one percent in the dust, and we're below OSHA in the air. What are you doing with your folks down there?" I asked Price.

"We got them suited up in Level C," he replied.

"Moon suits? At these concentrations?" I asked.

"Well, we just got this data, and we wanted the higher protection factor in case the levels went higher," he replied.

"What about respirators?" I asked.

"Half-face P100s right now," Price said.

I pressed him further: "What would you recommend for firefighters?"

Price had brought along his deputy, George Corcoran, a burly Bronx native and veteran emergency manager. Corcoran looked at me and gestured passionately, talking with his hands. "It's tough to say with this data. Ya gotta look at the big picture, though. You're probably talking P100s on those guys."

"Do we need that?" I said. "OSHA doesn't require them at these concentrations. I don't even know if the Fire Department has any."

"It's a tough call," Corcoran admitted, "but at our shop we have to be careful. All we got is one data point. It ain't the universe. But it could get worse down there."

Con Edison had ordered all of its employees working in and around the World Trade Center site to wear protective coveralls[146] and P100 respirators. The P100 is a molded face-piece with two cartridge filters that remove 99.97 percent of airborne particles. It is more protective but harder to wear and maintain. As it would do at any worksite, Con Edison was requiring maximum protection for its workers until it had data to prove that it wasn't needed.

I stared at the air sample result at the top of the page and struggled with the idea that we would recommend that FDNY pull its firefighters off the pile. How could we slow the rescue, with perhaps 15,000 people there, trapped in the rubble?

To put workers into P100s meant we would have to make sure they were physically capable of wearing them. They would have to be fit-tested and trained. And besides, how many P100s could we even get our hands on?

All of this based on a single data point—one that showed levels below the legal threshold, what OSHA called the per-missible exposure limit? A single data point that wouldn't be enough to require workers at a normal construction site to wear P100s?

It wasn't enough data to be statistically meaningful; not enough to draw conclusions about the conditions everywhere around an outdoor sixteen-acre construction site.

But this was not a normal construction site.

I called Ben Mojica, and we changed our safety advisory. From that moment, the New York City Department of Health recommended that all workers at Ground Zero wear P100 respirators.

Then I went to find some. I weaved over to Department of Citywide Administrative Services,[147] the city's procurement agency. It was five people huddled around a small desk talking into their cellphones and writing furiously. One of them looked up. It was Lisa Sacks, assistant commissioner.

"Lisa, I need ten thousand P100 half-face dual-cartridge respirators with replacement filter cartridges."

"Where and when?" she asked. "And oh, by the way, what is a P100 respirator?"

I pushed my way to the front of a line waiting to talk to OEM's shift commander, Richard Rotanz.

I told him what we were doing.

"What the hell is a P100 respirator?" Rotanz asked.

When I told him, he said, "How are they going to talk to each other?"

"They gotta yell," I replied. "I don't know, Richie, but the air tests I'm looking at tell me they have to wear them."

"Make it 20,000 then. You're gonna need 'em. You get them downtown. I'll let FDNY know they're coming."

We had taken the first step, but with hundreds of construction workers, first responders, and volunteers crawling over the remnants of the World Trade Center, underneath which fires still burned, we needed to figure out how to make it happen.

too hot to handle

Around noon, Rotanz steered the Environmental Protection Agency, in the form of its on-scene coordinator, Brett

Plough, over to me. Plough was the agency's first responder, a tall, thin forty-year-old engineer in a bright blue polo shirt with a red, white, and blue logo that read "EPA Emergency Response." I told him about Con Edison's air data and asked him for EPA's data.

Brett looked confused. "What do you mean?"

"What do you guys have?" I said. "Have you characterized the site?"

"No. *We* don't characterize *your* site," he said. "We are here to support the city. What is the city requesting at this time?"

"Look," I said, "We need to test for anything hazardous that could be in the air, all around the site."

"When do you want this done? And for how long?" he asked.

"How about around the clock from now until forever," I said.

"What does 'forever' mean?"

"Six weeks, six months, six years, who the hell knows?" I said. "Just get going and keep going."

A few minutes later he pushed through the crowd and gave me a thumbs-up sign. "I got your data. It's on its way."

An hour went by, but the data didn't show. That hour stretched into two. Still nothing.

I began to search harder everywhere else, looking for safety officers and their data. Surely there were more agencies, not just the EPA, on the scene collecting samples of the air and debris. Only data could tell us whether the air at the World Trade Center site was hazardous. How could we be thirty hours into this disaster and not have good data?

The afternoon sped by and every so often I'd hear a different excuse from Brett: EPA couldn't reach the right person, the sampling equipment had malfunctioned, or the samples were delayed getting to the lab. Every report ended the same way: "It will be here in an hour."

Finally, around 5 p.m., Plough arrived, waving a sheet of paper containing air data from one location in Jersey City, New Jersey, and two in Brooklyn. When I asked to see results from the World Trade Center site, or anything in Manhattan, he had nothing to show me.

"Oh," he said. "We didn't sample there."

What was going on with these guys?[148]

I was baffled. Based on my fifteen years of experience with the United States Environmental Protection Agency, I would have expected it to seize the reins of the investigation, calling all the shots. After all, it was the chief architect and protector of the Clean Air Act. The EPA had defined acceptable air quality everywhere in the United States and had been arresting violators for over thirty years. Now, the agency that literally wrote the book on air sampling didn't seem to know how to do it. As we were soon to discover however, somebody at the EPA was testing the air somewhere.

air that is safe to breathe

The next day, EPA Administrator Christine Todd Whitman announced that "short-term, low-level exposure of the type that might have been produced by the collapse of the World Trade Center buildings is unlikely to cause significant health effects." Whitman went on to express the agency's relief "that there appears to be no significant levels of asbestos dust in the air in New York City."

If the EPA had data to make that claim, it wasn't showing it to New York City. The rumor in the EOC was that the White House had directed the EPA to reassure New

Yorkers—those working at the World Trade Center site and those anxious to return to their homes and jobs downtown—to fast-track the recovery.

The impact was immediate. Our Health Department teams were already bringing the P100 order out into the field, handing out flyers during shift changes and explaining to workers the new requirement for P100s and how to get them. In the EOC, we began every report in the Daily Agency Briefing with a reminder of the P100 order. But after the announcement by the EPA administrator, FDNY and NYPD began to call us out in these meetings. They wanted to know why, if the EPA said the air was fine, our own Health Department was telling workers to wear these face masks. It was a good question.

We needed proof. Since nobody had it, and all the people who on a normal day were responsible for these things suddenly wouldn't touch them, we would have to go and get it.

into the red zone

The next day, a Health Department caravan loaded with environmental technicians and air-sampling equipment drove south through the empty streets of Lower Manhattan.

I had made an ask. I called an old friend, Alex Lempert, who was in charge of environmental issues at the NYC School Construction Authority. Lempert had huge air-sampling resources at his command. When I told him that I was going to test the air at the World Trade Center site and needed his help, he said, "Let's go. What do you need?"

It was late in the afternoon. We traveled through multiple checkpoints manned by NYPD and National Guard with machine guns at the ready. All along the way, buildings and sidewalks were covered in a thick coat of dust; stores were empty, frozen in time.

Finally, we reached the FDNY Incident Command post. Across the street was the pile, a hundred-foot-tall tangled mass of I-beams and concrete; stainless steel shards that used to be the exterior walls were tilted and askew amidst the debris. Tiny figures of firefighters were scattered across the face of the mountain.

I stepped out of the van and a teenage girl appeared out of nowhere wearing shorts, flip-flops, and a T-shirt. This girl had apparently evaded NYPD and its machine guns. Without a word she reached into a white five-gallon bucket and offered me a sandwich in a Ziploc bag with a red, white, and blue ribbon that read "God Bless You."

I looked around and saw other visitors. In addition to police officers, firefighters, and soldiers doing their jobs, there were people from random federal agencies like the Department of the Interior, Commerce, and Veterans Affairs, all of whom happened to be in the city on unrelated business and had used their government-issued IDs to get downtown. They stood around gazing at the surreal scene, doing nothing except to put themselves in harm's way.

The FDNY Incident Command post was a collection of brightly lit white tents in the middle of the intersection at West and Vesey Streets. I waded into the crowd to present myself to the Incident Commander. When I found him, he was standing in front of a whiteboard with a hand-drawn map showing the World Trade Center site as a grid of sixteen one-acre quadrants, with a battalion chief assigned to each one. I started to introduce myself, but he cut me off. He knew who I was and why I was there. He walked away, and I stood there, waiting…

Then, one by one, the battalion chiefs appeared.

They listened politely as we described what we were doing and then stood stock-still—they had that *thousand-yard stare*, the blank, unfocused gaze of soldiers who have become emotionally detached from the horrors around them—as we strapped the air samplers, each about the size of an old portable tape player, onto their waist belts and clipped intake tubes to their collars. Then they disappeared back into the pile.

We strapped samplers to everybody we could: construction workers, crane operators, police officers, and FBI agents. We took the rest of the samplers we'd brought and taped them to streetlight poles and fences, anywhere we could pull a decent sample.

At the same time, the formal requests that I, and others at the Health Department, had made for air sampling were starting to bear fruit. Over the next two weeks, OSHA would collect more than 250 samples for asbestos and began sampling for silica; carbon monoxide; heavy metals like lead, cadmium, mercury, and arsenic; and volatile organic compounds like benzene, toluene, and formaldehyde. The City's Department of Environmental Protection had set up testing stations all around the site and was collecting hundreds of air samples.

The data we collected that day convinced us that P100 respirators needed to be worn at the World Trade Center site. Unfortunately, it wasn't enough to convince the people who needed to wear them.

plugging into the great machine

Throughout those long early days, dozens of experts from an alphabet soup of city, state and federal agencies, including the Department of Health and Human Services and the Centers for Disease Control (in particular its National

Institute of Occupational Safety and Health, or NIOSH) had come together at the Department of Health's temporary headquarters on First Avenue. They had compiled a laundry list of public health and worker health issues that needed to be immediately addressed at the World Trade Center site.

So, we dispatched a separate team to the EOC on a mission to create order out of the chaos. We were assigned to work with Sam Benson, OEM's Health and Medical Director to convene all the organizations who were working on health and safety issues at the site, including FDNY and NYPD, the city's Department of Design and Construction, Port Authority, Bovis and the Operating Engineers, Bechtel, CDC/ NIOSH, OSHA, EPA, Deparment of Sanitation, FBI, and many others. We came together as a team—in a dingy upstairs conference room we called the "martini sky lounge"—and dubbed ourselves the WTC Health and Safety Task Force. And we kept the mission front and center in everything we did: *protect workers and first responders and improve health and safety conditions on the ground at the World Trade Center site.*

We worked around the clock,[149] coming together twice a day to solve problems and to gauge our progress. In the beginning, the meetings were raucous. We struggled to keep order as the standing-room-only crowd, that included construction bosses, Army colonels in fatigues, FDNY chiefs in uniform, congresspeople, and agency heads in business suits, all trying to comprehend and contribute as we worked through our list of tasks and problems. The meetings started big and got bigger, and within a week we went from a handful of data points to hard drives full of data. Information management quickly became a huge job, with dozens of Health Department staff vetting, analyzing, and presenting the data to us.

In the early days, immediate life safety was our biggest concern. Hundreds of people worked on the pile day and night using cranes, excavators, and grapplers to remove the mountain of steel and debris. Firefighters worked alongside the heavy equipment, manually digging through unstable mounds of debris and climbing down into the smoking voids. The risk of serious injury or death was extreme. The World Trade Center site was, in the words of OSHA Administrator John Henshaw, "the most dangerous workplace in the United States."

At the same time, we kept a laser focus on the hazards in the air. We sent roving teams of safety inspectors in "Gator" vehicles to patrol the site. One of their main jobs was to make workers wear respirators.

When we heard that workers were confused about where respirators were required, we posted maps showing a perimeter around the site (the famous "green line") within which respirators had to be worn. We even painted a green line on the streets and sidewalks. We established a dust-suppression program that included a nonstop spray of water on the pile and wash-down stations for workers and vehicles.

We worked to make sure respirators were available everywhere and at all times. We dispatched teams to conduct fit-testing. By September 16, we had a formalized fit-checking schedule in place at various locations near the site staffed around the clock with doctors and nurses conducting medical clearances for fit tests.[150]

We mounted a massive safety education campaign with official advisories, pamphlets, laminated cards, and huge signs affixed to every entry location at the site. We worked with the Operating Engineers[151] to create a mandatory eight-hour safety course that educated workers on the hazards at the World Trade Center site and what they needed to do to protect themselves.[152]

three strikes and you're still working on the pile

But the missing piece was enforcement; we needed a way to enforce the respirator requirement on the pile. Some have questioned the need for this. They say that the individual first responders and construction workers were responsible for keeping themselves safe by wearing their respirators. But they couldn't be more wrong. The first responders and construction workers were selflessly dedicated to a solemn mission and, like everyone else, they were confused by mixed signals.

"If the EPA says the air is safe, why do I have to wear this face mask?"

"Nearly every person around me, every police officer and firefighter and construction worker, has a mask hanging around their neck instead of over their mouth and nose. Why should I be the odd person out?"

"I see OSHA and the New York State Department of Labor here. If the rules require all of us to wear these face masks, why is nobody getting kicked off the site?"

From the beginning, we knew that we needed a three-strikes rule that would permanently bar workers from the site for refusing to wear a respirator. We knew this worked, because it was working all day long on Staten Island.

Every day, debris from Ground Zero was trucked from the site through the Brooklyn Battery Tunnel into Brooklyn and across the Verrazano-Narrows Bridge. Once it arrived at the former Fresh Kills landfill site on Staten Island, NYPD oversaw sifting through the debris for evidence and human remains. Deputy Inspector James Luongo was the Incident Commander. At its peak, as many as nine hundred law enforcement officers worked around the clock there.

Luongo did not need anybody to help him enforce the rules. Anyone caught not wearing a respirator was immediately relieved of his or her duties and kicked off the site permanently. There were no exceptions.

"Ladies and gentlemen, this ain't baseball," Luongo liked to say. "In my game, you only get one strike."

Those debris trucks hauled nearly two million tons of contaminated World Trade Center dust and debris off the site and through city streets. Although the Department of Health couldn't enforce worker safety regulations, we had broad authority to enforce our own rules that prohibited trucks from contaminating the tunnels, bridges, and streets along the route.

We used that authority to issue hundreds of tickets (each with a 500-dollar fine) to debris trucks that left the site unsealed or unwashed. This earned me the dubious distinction as "the most hated man at Ground Zero." But it worked. We enforced a zero-tolerance policy for contaminated World Trade Center dust on the streets of New York City. Unfortunately, as we shall see, we couldn't use health inspectors to issue tickets to workers.

Despite a massive effort, we could not force all the first responders and construction workers to wear a P100 respirator at all times at the World Trade Center site. We needed a zero-tolerance policy to keep them safe. We saw it working on Staten Island and on the streets around the WTC site and knew that it would work on the pile.

But very early on, at the highest levels of government, a policy decision was made: *no enforcement.* This no-enforcement policy is to blame for failing to protect the first responders and construction workers at the World Trade Center site.

going native

Legislators create laws, and agencies write regulations that implement the authority of those laws. Those regulations are enforced by the agencies that write them and only the agency that authored a regulation can enforce it. In the case of Ground Zero, OSHA and the New York State Department of Labor were responsible for worker safety; the New York State Department of Labor wrote the rules for public sector workers, such as firefighters and police officers, and OSHA wrote them for everyone else.

OSHA and the New York State Department of Labor would not conduct "enforcement activities" at the World Trade Center site, despite repeated requests by the city of New York they do so. The members of the WTC Health and Safety Task Force were frustrated. We called out the OSHA and New York state representatives at every safety briefing and demanded that they justify their failed policies.

They were all there: EPA, OSHA, and the New York State Department of Labor Public Employee Safety and Health Bureau (PESH) staff and supervisors had been with us, on the ground and in the EOC, from the first day. They were worked as hard as we did, and they cared.

But they could not explain, only apologize. OSHA, for instance, had been directed to "pursue collaboration while suspending enforcement…"[153] Their decision-makers, high up on the federal org chart, wanted to be involved, but they were desperate to avoid any mistake for which they could be blamed. These decision-makers stayed away from the safety briefings, so we tracked them down. One said to me, off the record: "We were told, in no uncertain terms, 'Don't go native.'"

For New York City and the WTC Health and Safety Task Force, and especially for the police officers and firefighters and construction workers in and around the World Trade Center site, this was not good enough.

the chicken is involved in a bacon and egg breakfast, but the pig commits

In every disaster, the ultimate authority is the top elected official, be it the mayor, governor, or president. In the disaster business, this person is affectionately referred to as "the Boss."

It would be easy to lay the 9/11 worker health failures at the feet of Mayor Giuliani, Governor Pataki, or even President Bush. After all, as the Bosses, they bore the ultimate responsibility. But they too were confused by the mixed signals from their own experts. No one expected them to come to our safety meetings. In a disaster, you can't expect the Boss to find the problem; you have to bring the problem (and the solution) to the Boss. This is the job of the Great Machine.

The Great Machine connects everything at all levels of the response. First, it connects agencies together in the EOC. Then these same agencies connect down to the boots on the ground and up to the Bosses.

This last part is critical because in the parallel universe, everything that can go wrong will go wrong all at the same time. We need the Bosses to quickly make the decisions that only they can make and to resolve the issues that only they can resolve; otherwise, we cannot act in the moment.

The local government did its job. OEM activated the ICS organization, the Great Machine, for everyone to plug into. But because everybody didn't plug into it, the failed policy couldn't get fixed. The WTC Health and Safety Task Force was a team within the Great Machine. All the issues of worker safety bubbled up into the Task Force, and we solved most at that level. But we couldn't solve the problem of the failed enforcement policy. So, we bubbled it up through the Great Machine to the Bosses. In this case, the mayor could have demanded an explanation from the EPA, the New York State Department of Labor, and OSHA—except that the USEPA Administrator, NYS Labor Commissioner, and US Secretary of Labor were nowhere to be found. They wanted to be involved with the World Trade Center response but would not commit.

bringing chaos to the chaos

Mistakes and blunders are pervasive in the parallel universe. Fortunately, we have an organizing system. ICS is the toolbox that disaster professionals bring into the parallel universe, and the Great Machine is the most important tool in that toolbox.

As a nation, we have not succeeded in bringing our toolbox to bear to minimize the blunders and to bring order to the chaos. The biggest obstacle to this is the federal government itself. Although it preaches[154] ICS, it doesn't follow it in practice.

September 11 is a good example of this. In the aftermath of the attacks in New York, FEMA, New York State, and New York City set up in separate locations. FEMA set up a Joint Field Office in another passenger ship pier south of Pier 92 on the Hudson River. New York State stayed in the State Operations Center 150 miles north in Albany. So instead of one disaster response, we had three. This dysfunctional environment enabled a failed policy that resulted in needless suffering.

This pattern—multiple agencies with different missions working separately with no accountability—is repeated in every disaster zone. Some might say, "That's not fair; 9/11 was almost twenty years ago. We have made progress since then." And to that I would say that, with respect to ICS, nothing has changed.

blackout

Fast-forward to the last week of November 2012, to the New York City EOC as, far out in the Atlantic, Hurricane Sandy makes a hard, left-hand turn and heads straight for the Eastern Seaboard.

For over ten years, since the early days of the Coastal Storm Plan, OEM had been working closely with the National Hurricane Center. Like our colleagues from across the East Coast and Gulf States, we had traveled to the hurricane center in Miami to train for the days and hours prior to a hurricane strike.

The process was logical, consistent, and crystal clear. We drilled it constantly and activated it in real time for every hurricane. It worked well in August 2011 in the run-up to Hurricane Irene.

This year, in the run-up to Superstorm Sandy, everything changed.

Hurricanes get their ferocious energy from the latent heat of the ocean waters, sucking up heat energy from the water like liquid through a straw. Sandy came late in the hurricane season. Since the ocean waters were cooler, Sandy was not forecast to become a hurricane but instead expected to turn "extratropical" before landfall. Even though tropical storm conditions and hurricane-force winds were inevitable, Sandy would not technically be a tropical cyclone, and thus would not fall within the National Hurricane Center's "jurisdiction."

It was decided at the highest levels of the vast federal bureaucracy that the National Hurricane Center, the foundation of our logical, consistent, and crystal-clear process, would go "dark."

The new plan was for emergency managers up and down the Eastern Seaboard to work with a new team of experts from a place called the Hydrometeorological Prediction Center.

The problem is that the Hydrometeorological Prediction Center wouldn't talk to us about the one thing we wanted to talk about. Storm surge, a fast-moving wall of water that drowns people, is the primary hazard from hurricanes, and causes more than half of all deaths from hurricanes. Because of this, the Coastal Storm Plan calls for a mandatory evacuation when significant storm surge is predicted. New York City made that call the year before when the Mayor ordered a mandatory evacuation. In August 2011, nine thousand patients were evacuated from hospitals and nursing homes in coastal areas before Hurricane Irene made landfall.

When we heard that the National Hurricane Center would not be involved, we were baffled. We reached out to them, and to everybody else we could think of, using every contact in the OEM rolodex. We wanted to talk about what Sandy's extratropical characteristics meant for our storm surge predictions. The experts we tracked down would not discuss it, except to say that there was a 'blackout' on the issue because "only the Hurricane Center can talk about storm surge."

The local National Weather Service office assigned a forecaster to the Emergency Operations Center. Day after day in the run-up to the storm, we gathered everyone, every 4 hours around the clock, to review our evacuation decision. And, every 4 hours, we talked about storm surge, while the Weather Service forecaster would talk about a 'potentially deadly storm' with high winds and flooding.

The problem was that in New York City, any random thunderstorm could be 'potentially deadly' with high winds and flooding. We needed *numbers*, specifically we needed predicted height of water *above ground level*, or AGL. Nothing less than forecasted AGL data would work for our evacuation decision. Unfortunately, we soon realized that the National Weather Service forecasters, inside and outside the EOC "did not clearly understand what storm surge was or how dangerous it could be."[155]

And there was no way to fix it, no Joe Lhota to call, no way to turn the gears in the vast machinery of the federal organization chart. Finally, late one night, we got a phone call from Donald Cresitello, hurricane program manager at the US Army Corps of Engineers.

"The hurricane center wants to talk to you," he said. The experts at the National Hurricane Center had taken it upon themselves to defy the failed blackout policy. Despite orders to the contrary, they reached out to give us the advice and the AGL data we needed to make the right decision.

bringing chaos to the chaos

Those who think we have made progress in our ability to come together as a nation during disasters are mistaken. We know this because we continue to see the same patterns of dysfunction in every disaster. Young staffers new to the EOC are often startled when they encounter it. In addition to setting up in locations far removed from the epicenter of the response, the federal government during catastrophes mirrors the daily workings of the bureaucracy itself, bringing forth an invasion of undisciplined and politicized agencies, headed by powerful secretaries that stride the earth like giants, accountable to no one.

From the local perspective, there seems to be little accountability anywhere. The middle-management layer on the federal org chart has been conditioned to distrust state and local governments with a strong "don't go native" culture. Therefore, federal agencies don't plug into anybody's Great Machine.

The result of this dysfunctional culture is a massive missed opportunity. Our federal government has enormous assets and talent. It has the potential to convene the greatest of Great Machines.

If we go back in time, nearly fifty years, to that disastrous wildfire season in California, we can review the problems that were brought to light: issues like lack of accountability, poor communication, lack of interagency integration, and freelancing. It is these problems, the very ones that prompted FIRESCOPE to develop ICS in the first place, that federal agencies continue to bring into the field and will bring to the next catastrophe in this country. Dozens of high-level federal officials will deploy without authority and they will freelance, bringing chaos to the chaos.

8 hungry ghosts

it's about the people, stupid

"The mission of humanitarian aid is to save lives, alleviate suffering, and maintain human dignity."[156]

—European Humanitarian Action Partnership

home fire part 1 | *Early morning* | *Sunday* |
Brownsville | *Brooklyn, New York City*

"Fire officials said the fire started around 4:30 a.m. in the fourth-floor walkup near East 98th Street and Sutter Avenue. Two single mothers lived there with their five children. They said there was no warning before the fire broke out. Fire crews scrambled, but with flames engulfing the apartment by the time they arrived, a lifetime of possessions melted and burned in a matter of minutes.

'We have nothing but the clothes on our backs. That's all we got,' said fire victim Teresa C."

Not only is Teresa C. homeless, she is desperate. Last night, in those dark hours just before the dawn, something took hold of her. It was the crisis, reaching out and pulling her, and her family, backward into the parallel universe.

Now she is trapped there, in a place of chaos and fear. She can't think clearly. Even simple questions like, "Do you want something to drink?" or, "Are you cold?" don't have easy answers. She has lost hope. She doesn't believe things will ever get better for her.

We can't really understand what she is going through, because we are in this world, looking across at her, alone in her parallel universe. We are not sure what we can do for her, or even what to say.

It's as if Teresa C. and her children are lepers. The police officers are polite to her, and the firefighters are nice to the kids. But they don't get too close, for fear of getting sucked in.

people in disasters need someone to help them

Teresa C. is paralyzed. She's no good to herself or her children right now. She needs someone else to help her.

Having spent so much time in the parallel universe during the post-9/11 period, in New York City we know a few things about it. We know what it means to work there and how hard it is to get things done. We know that excuses are worthless and that we are judged by what we do (and don't do) in the first hours of the crisis.

But those lessons apply to neighborhoods and cities. The challenge when providing humanitarian aid to children and families in crisis is even harder. The only thing that matters then is what we do in the first minutes.

every person needs a different kind of help.

Teresa C. needs help now. But what exactly does she need?

As much as we want an easy answer to that question, there is no easy answer. Like every person, every disaster is unique. The need is dependent on the person, the nature of the crisis, where the person is in life, and where he or she is in the timeline of the crisis.

"This is too complicated," you may be thinking. "Can't we just give money?"

Well, everybody always needs money, especially if it's free. But the real need is not always just money. And, as we have seen with FEMA, free money is a pernicious thing. It can create as many problems as it solves.

Humanitarian aid is about understanding what will get people on the road back from the parallel universe. Some people need a car; others just need cab fare. Some people need long-term counseling; some, just a compassionate word.

Some people need a roadside-Buddhist-hungry ghost ceremony.

world wide tours disaster

The Mohegan Sun hotel and casino is a gleaming silver-and-glass monolith towering high above the Connecticut forests. Its gaming floor, one of the largest in the country, is eight acres of busy patterned carpets and mesmerizing lights. It is here that a New York City subculture of overnight gamblers plays the slots and tables for a few hours before catching a late bus home. Most sleep during the trip; some arrive back just in time to return to work. It was on one of those late-night rides back to the city that the crisis suddenly and brutally struck.

mass-fatality incident

The top of the bright morning sun was just appearing over the horizon as a World Wide Tours bus, half full of gamblers, traveled south on the New York State Thruway en route from Mohegan Sun to Manhattan's Chinatown. As it crossed the northern border of New York City and entered the Bronx, it veered suddenly off the road.

The bus driver claimed that he was cut off by a tractor-trailer as it passed him on the left, changing lanes too soon and clipping his front bumper. We do know that the bus driver, Ophadell Williams, had been awake for most of the past three days and was barreling at over eighty miles per hour down a particularly treacherous stretch of highway.

As he lost control of the bus, it toppled and skidded on its side in a screeching shower of sparks.[157] It traveled nearly 150 yards along the right guardrail before slamming into a highway signpost. The heavy-duty steel column ripped through the front window, shearing the bus in half from front to back at the passenger window line.

The penetration of the signpost deep into the carriage caused a grisly scene inside the doomed bus with mutilated passengers—some decapitated, some hanging upside down, screaming in the darkness and struggling to get out. Some were thrown out onto the roadway, while others were trapped inside the ruined maze of metal.

> "'It was a pile of humans, wrapped in the metal, wrapped in the wreckage,' said Capt. James Ellson, 42, a twenty year veteran of rescues and fires who was among the first on the scene. '...from the front [of the bus] to the rear there were bodies. It was just a pile.'"[158]
>
> —Robert D. McFadden, *The New York Times*, 12 March 2011

Fifteen of the thirty-three people on board were dead, many of the survivors were critically injured.

Every crisis is an aberration, unique in every way, but these are the worst. As its name implies, a mass fatality incident is a surge in human death. Every one is a catastrophe: for the victims and their families, for their community, for the city, and for the nation.

Nobody escapes the brutality of it; not even the professionals we think are immune. Those who are called to respond to the scene are the first witnesses to the carnage. Few walk away unaffected.

Although disaster professionals don't witness them, we nonetheless fear them. In the early hours, as helicopters hover over firefighters and police officers working amidst the devastation at the scene, we must work in the tail of the response to place a laser-like focus on the survivors.

We must travel into the parallel universe and connect with them. We must stand beside them as they confront the crisis. This is grief of a uniquely intense and painful variety, the kind of grief that can come only from the sudden loss of a vibrant loved one. Their mother, father, sister or brother was expecting to see them. Now, they never will. Many of the survivors would give anything—even their very lives—to be able to say goodbye, to tell them they loved them.

These are the most desperate moments of their lives. Some are rendered helpless, completely ignoring their own needs—such as for food, sleep, or showers—driven by an unquenchable thirst for answers: "Where are they now? How did they die? What were their final moments like?" Only a constant stream of information seems to help, and no detail is too small.

In the aftermath of a mass fatality incident, disaster professionals must wade through the grief and chaos to tackle the surge of issues, none of which can wait: to provide that food, sleep, and showers; to site, staff, equip, and supply a reception center; deal with hyper-aggressive media; schedule police interviews and family briefings; support investigations at the scene; arrange for transportation and coroner interviews; and coordinate site restoration and formal death notifications.

family assistance

On that Saturday morning, OEM ran into the Situation Room and turned on the Great Machine. We plugged into the local elected officials and others who could help us. Within a few hours, we had set up a Family Assistance Center with a local community group, the Chinese Consolidated Benevolent Association.

Pictures from the devastated crash site filled the airwaves, and the city, especially the close-knit neighborhood of Chinatown, was deeply affected. The Family Assistance Center opened at the Chinese Community Center on Mott Street that same day, staffed by representatives from government agencies, including the Departments of Health and Social Services and disaster relief organizations such as the Red Cross.

Because this was an "open-manifest" job,[159] we didn't know who the families were, and so we worked with local media and social media to push the location of the Family Assistance Center far and wide. We wanted anyone who knew, or suspected, that a family member could have been aboard that bus to come forward.

Over the course of that day, the families did come, many because they didn't know what else to do. They crowded into the Chinese Community Center auditorium. Some stood around in stunned silence, while others talked with elected officials and Chinese Consolidated Benevolent Association staff.

Around the perimeter of the room were tables with the banners of the city agencies and faith-based and volunteer organizations who were there to offer services such as financial assistance, housing, legal services, immigration assistance, insurance advocacy, and assistance with animal care.[160] It got so busy on that first day that the Duty Team staff in the Situation Room worked to complete the planning for an extended operation lasting ten days or more.

By the next day, something had changed. The auditorium was quiet. Families attended the formal briefings but didn't return to the auditorium. The Family Assistance Center staff was left talking mostly to themselves.

Those people who did come kept asking to go to the scene. This is not an unusual request in these situations, but it was a perilous site, alongside a busy stretch of highway cut through a hillside and lying at the bottom of a steep curve. We discussed the idea and decided the logistics around it were too complex. We were not sure we could keep the families safe, so we told the Family Assistance Center to stay the course. In the meantime, I braved the crowded maze of Chinatown streets to find out what was going on.

The Chinese Community Center "auditorium" was a large conference room with dreary linoleum floors and a low white tile ceiling. There were a few family members there when I arrived, but I could not make any headway with them. After a while, I approached a pleasant-looking group in navy blue sweatshirts with bright white collars. These were the Tzu Chi volunteers. We had worked with Tzu Chi, a Buddhist international humanitarian relief agency based in

Taiwan, many times in the past. It's popular in the Chinese community because it doesn't require a lot of paperwork and because it gives cash.

Tzu Chi had spent many hours with the families, and one of the volunteers, Henry Chu—a tall and distinguished Hong Kong native—took me aside. He said that the mothers were distressed, not just because their children had died, but about the *way* they had died. According to Buddhist tradition, the intensity of the crash had damaged their souls. The sudden violence of their final moments had turned the spirits into "hungry ghosts"—strange beasts driven by intense emotions and condemned to roam the place of their death. According to the Buddhist monks who were with the families, the spirits of their loved ones were trapped on the roadside and only a ritual funeral ceremony could release them.

He then told me what we already knew, that the staff at the Family Assistance Center were baffled by this. In much the same way, the families were baffled by the "help" at the family assistance center, by the forms and the small-business loans, by the ill-fitting track suits, and the pamphlets filled with advice that meant nothing to them.

"Keeping the Family Assistance Center open seems like a waste of time at this point," Henry said, "There is nothing here for the families. The only thing they want is back there, on the Thruway."

the beast

As disaster professionals, we should not have been surprised. Anyone who has worked a mass-fatality incident knows the hungry ghosts. You can almost feel them, especially in the early hours when the loss is new. I have come to think of the grief as a huge, amorphous beast, looming over everything.

During these incidents, we focus on the families, because we have learned that—as strange as it may sound—the aftermath of a mass-fatality incident is not about the dead; *it's about the living*[161]. They are the true mission. Often, we get caught up in the complex choreography at the crash site or the media briefings and neglect the families. That is when we lose the job.

We use the family assistance center to help our focus on the families. Designed to be a single point of access to everything that government, nonprofits, and faith-based organizations offer, the family assistance center is one of the most important tools in an emergency manager's toolbox. But it's a one-size-fits-all tool and, although a good concept, it doesn't always work so well in practice. This is because there is no one-size-fits-all mass-fatality incident. Every mass-fatality incident, and every family who has lost a loved one, is completely unique.

The Chinatown families wanted only one thing from us. To give it to them, we needed to break a few rules.[162] My colleague Dina Maniotis and I turned to the Great Machine, and we made a plan.

Since the bus crash started in Yonkers—which is under New York State Police jurisdiction—and ended just over the New York City border, neither law enforcement agency would agree to help on its own. Once we convinced City Hall that we could make it happen and keep the families safe, NYPD was all-in. Once NYPD is all-in on a project, things move quickly. It worked with the New York State Police to organize a presidential-class motorcade to get us and the families to the scene.

Early one morning, exactly one week after the crash, the families and their spiritual advisors (along with OEM, Tzu

Chi, and the Red Cross) climbed into two MTA buses to begin the journey.

caravan

If you've never had a chance to travel in a motorcade, you really should try it. Patrol cars led the procession, followed by vans, then our buses, and then more vans, more patrol cars, and so on. We blew through every red light from the heart of Chinatown to the northernmost tip of New York City. With highway cops on motorcycles zooming by to block the intersection ahead, we did not stop once. Nobody in the history of the world has made it from Canal Street to the Yonkers border as fast as we did that Saturday morning.

The motorcade slowly rolled to a stop on the western shoulder of the New York State Thruway, a hundred yards south of the crash site. After a few minutes, the families emerged into the bright morning sun and slowly made their way to the spot near the highway signpost where the bus had come to rest. There the monks, in flowing orange robes and with shaved heads, began an elaborate funeral ceremony. They burned incense, chanted, and sang. Cymbals crashed. The families waved white and blue and red prayer flags and made offerings of pieces of cloth and plates of food.

The New York State Police had carved out the room for this by blocking a lane on the highway, and so the southbound traffic slowed to a crawl. As the drivers got a glimpse of the ceremony, they slowed even more. The spectacle soon slowed the northbound lanes too, turning them into a parking lot filled with rubberneckers, staring through their windshields at the amazing scene.

one size never fits all

As a one-stop shop, the family assistance center is a powerful weapon in any disaster professional's battle with the crisis. But in this case, the one-size-fits-all tool did not fit the Chinatown families. Their need did not fit the help we were offering. We kept wanting to give them the assistance they "qualified" for. The problem was that none of our programs had a "road-side-Buddhist-hungry ghost ceremony" benefit.

The problem is that, should you ever find yourself in the parallel universe, our one-size-fits-all disaster tools will not fit your needs either.

government help is one-size-fits-nobody

As an example, let's imagine that a catastrophe brings widespread disruption to your neighborhood. It could be anything—an earthquake, blackout, wildfire—you name it. In those dark days, as you and your family shelter in your damaged home, you hear a radio commercial with a toll-free number to call for FEMA disaster aid information.

When you call, you will engage the biggest and most commonly used tool in the disaster-aid toolbox: FEMA's Individuals and Households Program, or IHP. Unfortunately for you, IHP is a slow and convoluted tour through the worst sort of byzantine bureaucracy.

"Still Waiting for FEMA in Texas and Florida After Hurricanes

HOUSTON, October 22, 2017 (The New York Times)

*"Outside Rachel Roberts' house, a skeleton sits
on a chair next to the driveway, a skeleton child
on its lap, an empty cup in its hand and a sign
at its feet that reads 'Waiting on FEMA.'*

*Nearly two months after Hurricane Harvey made landfall
in Texas on Aug. 25, and six weeks after Hurricane Irma
hit Florida on Sept. 10, residents are still waiting for
FEMA payments, still fuming after the agency denied
their applications for assistance and still trying to
resolve glitches and disputes that have slowed and
complicated their ability to receive federal aid.*

*Brian Smith, whose home in the northern Houston
suburb of Kingwood had two feet of water inside
after Harvey, said, 'You feel abandoned.'*

*Rita Perreault, whose South Florida mobile home was
damaged by the flooded Imperial River, calls FEMA
twice a day to check on the status of her application
and inspection. Mrs. Perreault said she had spent so
many hours on the phone on hold that she learned, as
other callers have, to put the phone on speaker and go
about her day. 'I thought I was going to get brain cancer,'
Mrs. Perreault said. 'They give you the runaround.'"[163]*

It starts with housing. FEMA promises homeowners money to repair some (but never all) of the damage to their homes. But there are hoops aplenty ahead.

First, you must register with FEMA, fill out the forms, and produce a rejection letter from your insurance company. Eventually (up to ten days later), an inspector will come to your home to verify that it was in fact damaged. While the inspector is there, you must prove that you are who you say you are and that you own the home and that you live there. Then another ten or so days pass and a letter will come with FEMA's decision. For most applicants, the letter will say that the request for assistance is "denied."

Because the application requirements are so complicated, if you were denied assistance, you were probably just missing something simple like a signature or an old phone bill showing your address. Unlike most other government programs with a forms-based application process, FEMA doesn't tell you what you did wrong and how to resubmit an application. It just denies the request. At the bottom of the letter is the fine print that says something about how to appeal.

Now you're thinking, "I have a snowball's chance in hell of winning a government appeals process" or, "I would rather poke my eyeballs than subject myself to the frustration…" Even though you really need the help, you believe that resistance is futile. So, you give up.

What if you need more than just housing assistance after a disaster? You might need things like furniture, clothes, or transportation to work, or you might have disaster-related medical or dental bills. FEMA has a process for that, too. It's called Other Needs Assistance, or ONA. For you to receive ONA from FEMA, you must first apply for a small-business loan.[164]

I am not making this up.

"Applicants should complete the application, even if they don't want a loan. The SBA loan application is used to review an applicant's eligibility for additional assistance. For that reason, complete the application even if you don't plan to accept the loan."[165]

—FEMA Press Release after Texas floods, June 15, 2016

FEMA will send the small-business loan application form to you along with your housing denial letter. Like a lot of people, you would probably look at the form and think, "WTF, I'm not a small business" or, "My credit is terrible. It's a waste of time for me to apply for a loan when I am sure to be rejected" or, "My student loan payments are already sky high." So, you give up and toss the loan application form in the trash.[166]

When you don't apply for a small-business loan, the process immediately stops, and you become ineligible for any further assistance from FEMA. Believe it or not, you are not required to accept the small-business loan, just to apply for it. If you are denied the loan, you become eligible for Other Needs Assistance.[167]

Did I mention that Tzu Chi doesn't have paperwork?

mission: *maintain human dignity*

Humanitarian aid is assistance to people during disasters. Its objective is to save lives and alleviate suffering. Its overarching mission is to maintain human dignity.

Your tour through the worst sort of byzantine bureaucracy could only have been conceived by accountants in green eyeshades fearful of media reports of misspent disaster dollars. Not only is it not built for you, it's not built for anybody. The help that FEMA gives is always too little and too late. When it does come, it is given in a confusing and totally non-user-friendly way. If you are persistent, you might get some free money, but you will not come away feeling respected. You may not even come away with your dignity intact.

If the mission of humanitarian aid is to maintain human dignity, FEMA's IHP does not achieve the mission.

home fire part 2 | *Early morning* | *Sunday* |
Brownsville | *Brooklyn, New York City*

> *"Teresa C. said that she and another single mother
> and their children lived together inside the apartment.
> She scrambled to save her belongings as the fire
> fell through her ceiling. She ran to safety and then
> watched from a distance as the building collapsed.*
>
> *'We were raised in foster care, and we don't have any
> family,' she said. 'So, we are two women living together
> and raising our kids, and this is what happened to us
> today. I don't even know what to do. We have school
> uniforms for four kids and we have nothing.'*
>
> *Electronics and clothes that were pulled out to the
> sidewalk by the mothers were soaked and ruined
> from the water used to put out the flames.*
>
> *'We'll get through it together,' she said. 'And our
> children will be okay, because that's what's most
> important through all of it. I don't know how we're
> going to do it, but God will see us through.'"*

The job of the first responders is done. The Police Department kept them safe, the Fire Department rescued them from that burning apartment, and Emergency Medical Services treated those who were hurt.

Now what?

who owns them?

In August 2013, I left OEM to join the American Red Cross as Chief Disaster Officer for the Greater New York region. In Greater New York and around the nation, the Red Cross works in partnership with governments to respond to residential building fires. Red Cross volunteers are trained to "look for the job" (that is, find the people). They look for Teresa C. and her children standing on that cold sidewalk. Their job is to connect with her; to throw a lifeline into that parallel universe; first with a simple "hello" and eye contact, served with respect. They will say things like:

> *"I am so sorry for what happened; what you are going through right now…I'd like to get you and your children inside and talk about some ways I can try to help you if you are okay with that."*

Volunteers are empowered to provide support that includes food, housing, clothing, and rides to work and school. Most of all, their job is to give people hope that everything is going to be all right, so they can take that first step on the path to rebuilding their lives.

The Red Cross process is called case management and helps people like Teresa C., first by talking, then nudging, then gently pushing them toward the pathway back. And giving them what they need to take that first step.

For large-scale incidents that displace dozens of families at the same time, the Red Cross will open an emergency shelter in a local school or community center. People think a Red Cross shelter is just a place to sleep and eat, but I can tell you that it is much more complicated than that. People who seek shelter with the Red Cross are paralyzed. The mission of an emergency shelter must be to understand what will get everyone on the road back from the parallel universe.

One of the first things they do when they open a shelter is develop a closure plan. The closure plan isn't a logistics procedure for demobilization and cot dismantling. It's a spreadsheet with "head of household name" on the left and "recovery plan" on the right.

And we begin the recovery conversation with the heads of household immediately: "What is your plan? Where are you going? How are you going to get there? What do you need to get there?"

We don't judge. We don't try to fit people into a box. We just get them what they need. Whatever it is.

Different people need different things to get them off square one. It's not always just money, clothes, or housing (although those things are always needed). Sometimes we advocate for them with a landlord or a social service agency or a school principal. Sometimes they just need lunch.

the red cross owns them

Late one night, on a cold sidewalk in the Highbridge neighborhood of the Bronx, a veteran Red Cross volunteer lectured me on the state of the disaster business. With a smoldering apartment building as a backdrop, he concluded his speech with a flourish: "The public thinks that disasters are always about the collapsed buildings or the helicopters or the firetrucks. I always have to tell 'em, 'It ain't about any of those things. It's about the people, stupid.'"

Different organizations own different pieces of the disaster. The Fire Department owns the life safety piece. Emergency Medical Services and hospital emergency departments own the healthcare piece. The Police Department owns the security piece. Emergency management owns the critical infrastructure piece. FEMA owns the programs that give free money.

The Red Cross owns the people. I was fortunate to work with a dedicated group of professionals, volunteers, and staff who put a laser focus on the people; to get them what they needed when they needed it. For a long time, I thought we were the best humanitarian organization in the country at that critically important mission.

That is, until I met Tzu Chi.

the anti-FEMA

"We listen to them as they tell us what happened during the earthquake. We're not only providing aid, but also providing love to help rebuild their inner peace."[168]

—Tzu Chi USA volunteer, 2017 Central Mexico Earthquake, Tiahuac, Mexico

After the World Wide Tours disaster, we wanted to help the families of the victims, but we really didn't know what they needed.

Tzu Chi knew. In Chinese, "tzu" means "compassion" and "chi" means "relief."

New York City, especially its close-knit Chinese community, was deeply affected by the tragedy in March 2011. We sympathized with the families. At the same time, Tzu Chi showed compassion for them. Although we often conflate these concepts, they are not the same thing.

The word "sympathy" comes from the Greek "*sympatheias,*" or "with feeling." To sympathize means to feel sorry for someone regarding a loss or other misfortune. "Compassion" comes from the Latin "*compassio,*" or "to suffer with."

Every person in the parallel universe has a unique set of needs that depends on the person and the disaster. Tzu Chi knew what the Chinatown families needed because compassionate relief means traveling into the parallel universe to suffer alongside the survivors.

We have said that Tzu Chi is popular with people affected by disasters. At OEM, we thought it was because it doesn't have paperwork and because it hands out five-hundred-dollar cash cards. While this is true, it's not the whole story.

Let's imagine a scenario in which an investigative journalist with a hot lead calls the headquarters of Tzu Chi USA for a reaction to a breaking story:

> *"Our investigation has uncovered malfeasance on the part of Tzu Chi USA. Following anonymous leads, we tracked down instances of disaster victims' using 500-dollar cash cards provided by your organization to buy things like cigarettes and lottery tickets. In one egregious instance, a disaster victim used the card to purchase twenty-five cases of Bud Light beer. Your donor dollars are clearly being squandered. How would you respond to these allegations?"*

The Tzu Chi representative who answered the phone would be confused by the question and would respond with something like, "They should use the cards as bookmarks instead? Or to cut cake?"

Tzu Chi doesn't judge; it doesn't offer unwanted advice, doesn't tell you what you should be doing. It gives thanks for being allowed to serve. It gives cash with no strings attached. And it also gives something else, something much more valuable than money.

Tzu Chi gives respect.

9 invisible impacts

the disabilities and access and functional needs crisis

> "We have a responsibility...to protect our most vulnerable citizens: our children, seniors, people with disabilities. That is our moral obligation."
>
> —John Lynch, governor of New Hampshire, February 22, 2012

east harlem gas explosion

Legendary film director Sidney Lumet is known for gritty crime dramas that feature tortured characters in crisis, from Henry Fonda in *Twelve Angry Men* to Al Pacino in *Serpico* to Peter Finch in *Network*. Many of his movies are set in New York City. One of his best known is *Dog Day Afternoon*, starring Pacino and shot on location on Avenue P in Gravesend, Brooklyn.

One of Lumet's early works is *The Pawnbroker*, filmed in 1964 and starring Rod Steiger. It tells the story of a Holocaust survivor, Sol Nazerman, who lives in his pawnshop in Harlem. Nazerman's spiritual "death" in the concentration camps causes him to bury himself in the most dismal location that he can find: a slum in Upper Manhattan.

Much of the film was shot on location on upper Park Avenue in East Harlem. The building that housed Sol's pawnshop, 1642 Park Avenue, was demolished in the 1970s, and the lot sat vacant for decades. In 2009, a low-income-housing developer built a five-story apartment building on that same lot.

On a cold gray day in March 2014, gas from the labyrinth of steel pipes buried beneath the streets began to leak into the porous ground in front of that five-story apartment building at 1642 Park Avenue. From there it seeped into the basement of the building next door that housed the Spanish Christian Church on the first floor. All through that night and into the next day, the level of gas within the buildings continued to climb.

Although many people in the area later reported that they smelled gas, it was not until shortly after 9 a.m. on Wednesday, March 12, that anyone called the gas company, Con Edison. By then, it was too late. Eleven minutes after that first call came in, both buildings were gone.

At 9:31 a.m., a thunderous blast sent a towering plume of flames, smoke, and debris into the surrounding streets. Witnesses reported seeing people flying out of the windows as the buildings came down. The force of the blast shattered windows a block away and registered on the seismic scale at nearby Columbia University, which measures earthquakes in and around New York City.

As the buildings crashed down, people were trapped in their cars, in the rubble, and in neighboring apartments. Others made desperate rescue attempts, rushing toward the flames and smoke. FDNY firefighters in a firehouse five blocks away heard and felt the massive explosion. Within minutes, FDNY escalated the incident to a five-alarm fire, bringing over 250 firefighters to the scene. The FDNY also transmitted a 10-60 code over the radio, signifying a major emergency.

Morning television shows were preempted in favor of nonstop news coverage of the explosion and its aftermath. The toll was heavy: eight people were killed, seventy others injured, and many critically burned. In addition to the apartments destroyed in the two buildings, seven other buildings were damaged. A hundred families were instantly homeless.

the red cross in east harlem

As the Chief Disaster Officer for the American Red Cross in Greater New York, I kept close to my cellphone. On that Wednesday morning, I was conducting a job interview and determined not to be interrupted. As my phone rang and buzzed, the candidate, a retired NYPD sergeant, kept glancing at the television mounted on the wall behind my head. Finally, she said to me in a deadpan voice, "Do you think you ought to be en route to this job?" So I turned around to face the crisis.

Every crisis has a range of personality traits and among these is its trajectory. For the Red Cross, the East Harlem Gas Explosion disaster response followed a trajectory of high intensity from its first minutes.

We were listening to the OEM Duty Team call as we pulled up to the scene. OEM said that it was considering several reception center locations including Public School 57 on 115th Street, so we ran over and found the principal, Nancy Diaz. After we explained to her our desire to take over her school and fill it with desperate families, she did not hesitate. She called her staff over and they sprang into action, clearing out the second-floor gymnasium, the largest room in the school.

A few minutes later a security guard approached me, saying, "There's a young woman at the front door with a baby and a toddler. She says she lives in the building next door to the one that exploded."

Then they started to pour in, the mothers and children and seniors who lived in the nine buildings around 1642 Park Avenue. We set up a reception center where the affected families and friends could gather for mass care and emotional support. As the room filled up, we called for volunteers and supplies and food. As the word spread, more and more streamed in, and by midafternoon the tiny gymnasium was packed with over two hundred people.

These were people who had just run for their lives with only the clothes on their backs. There was little they didn't need. They needed prescription drugs and wheelchairs. They needed to find missing family members. They needed a roof over their heads. They needed someone to tell them that everything was going to be okay. They needed lunch.

Within a couple of hours, more than forty Red Cross volunteers and staff were there. The intensity kept building, and the afternoon passed in a blur. I remember running up and down those stairs about a hundred times.

At one point, I walked outside to take a phone call and into a bank of floodlights. NYPD was keeping a pack of media cameras behind a line in the sidewalk as politicians paraded slowly to the door.

Congressman Charles Rangel, who represented East Harlem at the time, turned to the cameras and said, "I've never had anything this horrific that's happened in my community since I've been in Washington…. It's our community's 9/11."

New York City Council Speaker Melissa Mark Viverito buttonholed me for 20 minutes to make sure I was doing everything possible to help the families.

We began to think ahead about the overnight, and around 4 p.m., while working with our partners at the Salvation Army, we started a process to open a shelter at its Manhattan Citadel just a few blocks away on East 125th Street. OEM sent MTA buses to transport the families. The process of getting everybody organized and down to the buses was a logistical operation akin to the evacuation of Dunkirk in 1941. At 11:30 p.m., the last person had settled into a seat and the buses pulled away. A fresh Red Cross/Salvation Army team was waiting to receive everyone at the Citadel.

A few of us stayed behind to close the reception center. As we walked into the empty gymnasium, I was startled by the sudden transition from chaos to quiet. Someone cracked a joke (something like, "Did anyone get the license plate of that truck that just hit me?") and we started to laugh. As relief washed over me, I dropped onto the hardwood floor in a fit of laughter. It was the best moment of my career.

We operated the shelter at the Manhattan Citadel for three days and two nights and, working closely with OEM, placed all the families from the destroyed apartments into transitional housing.

For the East Harlem Gas Explosion disaster response, nearly three hundred Red Crossers helped 215 adults and 113 children. We provided casework and financial assistance to over a hundred households, served nearly six thousand meals, and provided more than 1,300 emotional support contacts.

East Harlem gave me the privilege of being allowed to serve its people in the worst moments of their lives.

And then, a year later, it happened again.

East Village Gas Explosion

Exactly six miles, or 120 city blocks, south of 1642 Park Avenue is a spot known as "the mayor's residence." The building at 121 Second Avenue formerly housed New York City mayor Fernando Wood. An 1855 article in *The New York Times* describes a serenade by five hundred musicians taking place in front of that address.

In March 2015, 121 Second Avenue was a five-story brick tenement building in the East Village neighborhood of Manhattan. In the 1950s, the East Village had become an epicenter for artists and bohemians. Now it is known for its happening nightlife. Old-school bars, music venues, and performance spaces share the streets with posh cocktail lounges and hip restaurants.

On the afternoon of Tuesday, March 26, twenty-three-year-old Nicholas Figueroa was in the East Village. He was having lunch at Sushi Park, a restaurant on the first floor of that five-story brick tenement building at 121 Second Avenue.

At around 4 p.m., Figueroa and his lunch partner were ready to leave, so he got up to pay the bill. As he walked to the rear of the restaurant, two men nearly knocked him down running out of the restaurant.

The men were Michael Hrynenko of the Bronx, the son of the landlord, and Jerry Ioannidis, an unlicensed plumber whose rickety tangle of pipes and valves had caused a major gas leak in the rear of the building.

An hour before, a Sushi Park employee had smelled gas and called the landlord. When Hrynenko and Ioannidis walked in and saw the danger caused by their nightmare setup, they panicked. They were caught on surveillance tape "swiftly sprinting out of the restaurant without warning any of the patrons or workers."[169]

Moments later, a huge rumble sounded, and the lower facade of 121 Second Avenue slid out onto the sidewalk in a cascade of glass and loose bricks. The force of the explosion blew Figueroa's date out of the eatery and into the street. The ground shook, and debris, plaster, and glass flew through the air. Within minutes, flames were shooting out the front windows and towering through the roof. As the fire raged, there were desperate attempts to flee, with residents jumping from fire escapes, and dramatic rescues: off-duty firefighter Mike Shepherd, who happened to be in the area at the time of the blast, can be seen on video helping rescue a hysterical woman who was stranded on her fire escape. On the way down, the fire was so hot, it melted the soles of his shoes.

The fires were so intense that firefighters had to withdraw from the buildings and engage in a "defensive outside attack, pumping a deluge of water onto the structures."[170] Three buildings—119, 121, and 123 Second Avenue, on the northwest corner of East Seventh Street and Second Avenue—were reduced to rubble. An adjacent building, 125 Second Avenue, was severely damaged. Nicholas Figueroa was killed, along with twenty-seven-year-old Moises Ismael Locón Yac, a busboy at the Sushi Park restaurant. Nineteen people, including Hrynenko and Ioannidis, were injured, including four critically.

same job, different disaster

Our emergency communications center dispatcher heard FDNY call a 10-60 "major emergency" code over the radio, and we went immediately en route. A FDNY major emergency puts a tremendous amount of equipment and personnel into Manhattan's already crowded streets. I parked ten blocks south of the scene and race-walked to the command post. When I arrived, FDNY had sounded a seventh alarm and the command post was packed with agency personnel from FDNY and NYPD, Con Edison, the Office of Chief Medical Examiner, and the Department of Buildings. At that point, the focus was on rescue operations.

The Red Cross team started to work the job and worked it exactly as we had one year prior in East Harlem. We contacted OEM, who put us in touch with the principal at PS 63 on East Third Street, three blocks from the scene. We cleared out the first-floor gymnasium, the largest room in the school.

We called for volunteers and supplies and food. We set up a reception center where the affected families and friends could gather for mass care and emotional support. We spread the word. We knew that, as in East Harlem, there were people in the East Village who had just run for their lives with only the clothes on their backs who would need help.

Within a couple of hours, we had more than thirty-five Red Cross volunteers and staff in the makeshift reception center at PS 63.

At one point I walked outside to take a phone call and into a bank of floodlights. NYPD was keeping a pack of media cameras behind a line in the sidewalk.

But the families didn't arrive in those first minutes—and, even as the word spread, they kept not coming.

For the Red Cross, the East Village Gas Explosion disaster response and the East Harlem Gas Explosion disaster response were identical.

They had the same trajectory, same number of affected households, same urban density, same agencies, same cramped elementary schools, same number of volunteers.

Only one thing was different: in the East Village, there were no clients.

We knew that there were people in the East Village who needed prescription drugs and wheelchairs, a roof over their heads and lunch. There were people who needed someone to tell them that everything was going to be okay.

It's just that the people of the East Village didn't need the Red Cross to provide these things.

we're not in east harlem anymore

In my business, we sit around carpeted conference rooms talking about how the poor and underprivileged are an important part of the mission.

The neighborhood of East Harlem is a mere fifteen-minute ride on the number-6 subway train from the neighborhood of the East Village. But in terms of the proportion of poor and underprivileged residents, they couldn't be more different.

Fewer than one in five East Village residents are below the federal poverty level, compared to more than one-third of the residents of East Harlem. In fact, East Harlem is one of the poorest neighborhoods in the city. At thirty-three percent, the rate of obesity in East Harlem is over four times the rate in the East Village. The diabetes rate in East Harlem is thirteen percent, compared with three percent in the East Village. East Harlem adults have the third-highest rate of alcohol-related hospitalizations and the second-highest rate of drug-related hospitalizations in the city.

The East Village Gas Explosion disaster response taught me an important lesson: The poor and underprivileged are not an important part of the mission. *The poor and underprivileged* are *the mission.*

The problem is that I should have learned this lesson long ago, in the aftermath of Hurricane Irene, when New York City was taught a similar lesson.

bcid v. bloomberg

"Pandemonium did not reign; it poured."

—John Hendrick Bang

"ReeRee Rockette is on vacation in New Orleans with her boyfriend when Hurricane Katrina hits.

They are evacuated from their hotel, and the airport is closed. Standing on the street, soldiers point listlessly, and they take a long walk south, eventually arriving at the chaotic scene that is the Superdome.

As she makes her way in among a stunned crowd, she is told that there are no working phones.

Rumors are spreading that there has been a rape, a suicide, and a murder.

An Australian guy tells her it's not safe to be alone.

A soldier says the generator is about to fail, so they'll be in darkness and a riot may break out.

She is terrified.

Inside the Superdome, things deteriorated rapidly.

*Temperatures had reached the upper 80s,
and the punctured dome at once allowed
humidity in and trapped it there.*

*Food rotted inside the hundreds of
refrigerators and freezers spread throughout
the building; the smell was inescapable.*

In the bathrooms, every toilet had ceased to function.

*The water pumps had failed, and without
water pumps to the elevated building, they
couldn't maintain water pressure.*

Every sink was broken.

*The population of the festering, battered dome
had gone from 15,000 to 30,000 in a short time as
helicopters and vehicles capable of cutting through
the water picked up stranded citizens and brought
them to the only place left to go in the entire city.*

*"'We escaped the storm only to be moved into the
basketball arena with hundreds of sick and elderly. The
rumor is that after the sick, the foreigners will be next to
get out of New Orleans but, oh my God, the smell. Everyone
wears medical masks, but there are none left for us.'"[171]*

For nearly eight years, I was Deputy Commissioner at the OEM, where I directed disaster planning and led city-wide coordination for major disasters.

When I joined OEM in February 2006, the humiliations of the Katrina response were fresh in our minds, and on my first day at OEM, Commissioner Joseph Bruno called me into his office. "City Hall is fixated on the Katrina thing. They are horrified by what happened there."

"We all are, Commissioner," I said.

He handed me a white three-ring binder with a plain white cover that said simply, "Coastal Storm Plan."

"This is yours. We want the best hurricane program in the world. Nothing less."

So, in between managing the EOC for fires and blizzards and power blackouts, the New York City Coastal Storm Plan became our Job One.

We built teams and wrapped them around big problems like storm-monitoring and decision-making, debris removal, logistics, evacuation, and sheltering. We visited our counterparts in the Gulf States, including Texas and Florida, to learn their lessons and best practices and we researched everything we could find.

Shortly after I arrived, somebody sent me a link to a congressional report on the national response to the Katrina disaster. Over nearly four hundred pages, a Bipartisan Committee of the US House of Representatives described what was supposed to happen but didn't and why and who was responsible.[172] To this day, disaster professionals in New York consider "A Failure of Initiative" a masterpiece of disaster literature. At the New York City OEM, it became our bible for what *not* to do, and we used it to build the blueprint for the New York City Coastal Storm Plan.

Given the catastrophe at the Superdome and the challenges the Red Cross had faced across the Gulf States during Katrina, we decided that the city would take on the responsibility for mass care during disasters.

In partnership with many city agencies, including the Department of Homeless Services and the Department of Education, we built an emergency sheltering system with over five hundred emergency shelters. We developed an automated staff notification and deployment system and a shelter-staff-training program that included classroom and just-in-time field training. To supply the shelters, we created an emergency stockpile of medical and personal care kits, cots, blankets, food, water, and pet supplies that could sustain up to 70,000 people and could be deployed to shelters throughout the city within twenty-four hours.

We worked for five years to build this capability and then, in August 2011, we got to use it. As Hurricane Irene bore down on New York City, we activated the Coastal Storm Plan and all its moving parts. The mayor ordered a mandatory evacuation of coastal areas. About 350,000 people were ordered to evacuate.

We activated our shelter operations center and deployed the emergency supply stockpile. We opened over eighty shelters that, at one point, housed over ten thousand people, including over a thousand people with special medical needs. Five thousand city employees worked in the emergency shelter at some point during the storm.

Then, less than a month after we closed the shelters, the mayor of the city of New York was served with a subpoena. On September 26, 2011, the Brooklyn Center for Independence of the Disabled (BCID) filed a complaint in the Federal District Court in the Southern District of New York that alleged that Mayor Bloomberg and the city of New York discriminated against men, women, and children with disabilities by failing to include their unique needs in emergency planning.

According to the complaint, Tania Morales, a wheelchair-user living in Brooklyn, was turned away from an emergency shelter during Hurricane Irene. She said that this was because the gate for the ramp into the shelter was locked and the shelter staff could not find the key.[173]

The complaint alleged a "pattern of neglect" and "blatant disregard for the lives of persons with disabilities."

We were stunned. The emergency sheltering operation during Hurricane Irene was by no means flawless. The truth is that, as with every emergency sheltering operation, some parts were pretty ugly.

Managers, staff, and clients in every one of those eighty shelters encountered glitches and missteps from the first minutes, starting with getting the lights on to finding and unpacking supplies, setting up, registering clients, and on and on. Some staff had forgotten what they learned in training, couldn't find the shelter guides, and felt like they were forced to make it up as they went along. And there were more serious problems, such as sick clients and power failures. Staffers called in many of these issues to the shelter operations center that worked to resolve them one by one.

We looked closely at the glitches and missteps and concluded that, although the emergency shelter operation wasn't always pretty, we kept people safe. Thanks to the heroic effort of the shelter managers and staff and clients who pulled together, we considered the operation a success.[174]

And we could not substantiate Morales' allegations. Our internal investigation found no instances of anyone being turned away from any shelter for any reason. We went into defense mode and worked with our legal department to fight the case in court. People at dozens of agencies—including NYPD, FDNY, and the Health Department, nearly everybody involved in the planning and execution—were deposed. I spent several days giving sworn evidence under questioning by Disability Rights Advocates, or DRA, the plaintiff's lawyers.

The case went to trial in March 2013 at the federal courthouse at 500 Pearl Street in Lower Manhattan. Lawyers for the city argued that its emergency planning program was continuing to evolve and improve, while the DRA lawyers were adamant that it was not good enough. Then, Jesse Furman, federal judge for the Southern District of New York, issued his ruling.

We lost.

Judge Furman acknowledged the "Herculean task" that the city faced in planning for and responding to emergencies and disasters, and the city's extensive efforts to meet the needs of all residents in times of disaster, but...[175]

> "The question in this case...is not whether the City, or individual first responders, have done an admirable job in planning for, or responding to, disasters generally. They plainly have. Instead, the question is whether the City has done enough to provide people with disabilities meaningful access to its emergency preparedness program.... The answer to that question is that it has not."
>
> —BCID v Bloomberg, US, Southern District of New York, 11 Civ. 6690 (JMF) (2013)

The judge ordered the two sides to negotiate a settlement, and in September 2014, the parties announced a comprehensive agreement to remedy the deficiencies found by Judge Furman in his decision. The agreement required, among other things, that New York City improve its ability to evacuate persons with mobility disabilities from their homes, provide accessible transportation during evacuations, provide accessible shelters, and canvass areas without power for people who may be trapped.

Although not front-page news, the headline in the New York Post would probably read, "Judge Furman to the City of New York: You're Not Doing Enough."

The thing is, until BCID v. Bloomberg, we didn't know what "enough" was.

We had always considered the disabled and those with access and functional needs in our planning, but there was no clear standard toward which to build. We were doing the best we could with the time and resources we had, when Judge Furman drew a bright line for all to see. And because the standard is a federal court order, New York has no choice but to find the money to make it happen.

So, although we lost the case, it was in many respects a great victory—not only for the disabled and those with access and functional needs, and not only in New York. It wasn't until I left the city that I really understood. The lesson of BCID v. Bloomberg is that I, too, was guilty...*of arrogance.*

Of course, accommodating the disabled and those with access and functional needs in every aspect of disaster planning is hard. But it was my job. And guess what? Nobody asked me whether it was hard.

This becomes crystal clear when you work directly with people on the ground in the disaster zone. That is when you learn that the "Disabled and those with Access and Functional Needs," or DAFN, are not an important part of the mission. *DAFN* are *the mission.*

BCID v. Bloomberg was a landmark case. The settlement and the requirements set by it define the boundaries of the gap in our resilience. The US Department of Justice has moved aggressively to correct this gap. Its civil rights regulations require equal opportunity for DAFN in all aspects of emergency planning and response. Federal and state officials have issued thousands of pages of guidance that explain the new rules and what needs to be done.

But…we have a problem. The people who are responsible to actually do these things are struggling to figure out how.

Local disaster professionals are trying; they are convening support networks and developing plans. But local officials are already overworked and don't have the bandwidth or the money or the support they need to make it happen. The reality is that nobody can come close to doing what the rules require.

the disabilities and access and functional needs crisis

The other thing I learned at the Red Cross is that, instead of doing the kind of capacity-building for mass care during disasters that we did at New York City OEM, nearly every local government has abandoned the responsibility for mass care to the Red Cross.

At first glance, this makes sense. After all, the American Red Cross is one of the world's premier humanitarian organizations, with a mission to alleviate disaster-caused human suffering around the nation. It conducts large-scale field operations, such as emergency sheltering, feeding, and distribution of relief supplies, in the disaster zone. It is an important national asset with the right mission and thousands of dedicated volunteers and staff.

As I write this in the fall of 2017, after Hurricanes Harvey, Irma, and Maria and the California wildfires, the American Red Cross has wide-ranging relief operations in eight states, Puerto Rico, and the U.S. Virgin Islands. Twelve hundred Red Cross disaster workers are on the ground now in California, and over the past eight weeks, the Red Cross, has provided more than 1.3 million overnight stays in emergency shelters.

But, all of its great work notwithstanding, the Red Cross keeps not learning the lessons that can only be learned in the parallel universe, one of which is that success or failure is determined by what it does and does not do in the first hours of the crisis.

For instance, the Red Cross is baffled by the nearly universally held opinion of its performance during Hurricane Katrina as a failure. It insists that its volunteers worked hard during that response, running massive operations on the ground involving thousands of people. While all of this is true, it didn't do those things fast enough. Like Andrew before it and Sandy after, the Red Cross got a couple of days behind the job and never caught up.

As the Chief Disaster Officer for the American Red Cross in Greater New York, I was part of a dedicated team of professional volunteers and staff that responded to 2,500 residential fires and assisted ten thousand people every year. But that incredible team—fifty-five staff and three thousand volunteers, operations bases, warehouses, a twenty-four-hour emergency communications center, and an EOC—couldn't scratch the surface of what would be needed to bring humanitarian relief to all of the affected areas of New York in the early hours of a catastrophe. This is true of Red Cross disaster teams across this great nation. *Without exception.*

The American Red Cross does not have the resources it needs to deliver on its mission of humanitarian relief in the early hours of a catastrophe.

This is also true of every other nongovernmental organization, including the faith-based and voluntary organizations that help people after disasters. The problem is that if you asked your mayor, county executive, alderman or alderwoman, or city councilmember today who is responsible to deliver humanitarian relief in your neighborhood after a disaster, they will point to the American Red Cross.

So, across the nation, we are relying on organizations that can't do the job. A system that can't do what the public expects and what the law requires is a system in crisis. In September 2005, the world watched as the richest and most powerful nation on earth abandoned its citizens to the hellhole of the Superdome. The reality is that we are but one catastrophe away from the next Katrina Superdome.

part IV |
the way forward

10 a great machine for america

what you can do

"Determine that the thing can and shall be done, and then we shall find the way."

—Lincoln[176]

it's just not there

It is Tuesday, October 31, 2017, and as I write this, my city is reeling from yet another terror attack.

At 2:06 p.m. on a sunny and warm Halloween afternoon, a twenty-nine-year-old Uzbeki national rented a white Ford pickup truck at a Home Depot in Passaic, New Jersey. He headed west across the George Washington Bridge into Manhattan and then drove south down the West Side Highway.

He was on a mission, motivated by online propaganda such as a 2010 webzine article entitled "The Ultimate Mowing Machine," which instructed followers of Al Qaeda how to use a pickup truck as a "mowing machine, not to mow grass but to mow down the enemies of Allah."

At 3:04 p.m., he left the roadway and drove onto the Hudson River Greenway, a protected bike path that runs parallel to the West Side Highway. He proceeded south for nearly a mile, striking every jogger and cyclist on that part of the path, killing eight and injuring seven others.

He exited the bike path at Chambers Streets, just outside Stuyvesant High School in Lower Manhattan,[177] and crashed the truck into a yellow school bus, severely injuring four special-needs students before being shot by a New York City police officer and taken into custody.

In a press conference at police headquarters just minutes after the attack, New York governor Andrew Cuomo spoke.

"The truth is...New York is an international symbol of freedom and democracy. And we are proud of it. That also makes us a target for those people who oppose those concepts...."

In New York City, as in many other places, the threat of terrorism is real. New Yorkers worry about it, but we go about our everyday lives, because, as the governor said today: *"If we change our lives, we contort ourselves to them, then they win and we lose."*[178]

As a disaster professional, though, my worry about terrorism does affect my everyday life—because I, and my team, strive all day every day to make sure we are as prepared as we can be for this very real threat. Because of our commitment, and the commitment of my organization, I do not worry that NYU Langone Health is not prepared for that threat. For similar reasons, I do not worry that the city of New York is not prepared for that threat.

I do, however, worry that there are people out there who are much smarter than the person who perpetrated this attack today. There are people with greater intent, more resources, and better capability who also have us in their crosshairs. These people could perpetrate an attack on a scale that would overwhelm even the considerable capability of New York City and New York state, and require a response from around the nation and even around the world.

I worry about the impacts of that attack, the deaths, and the chaos. But perhaps even more than the incident itself, I worry about the aftermath, and about our ability as a nation to respond to it. After working in the disaster business for nearly twenty years, I know that, as I sit here today, there is no plan or program or ability to bring the nation together in support of this city in the aftermath of a catastrophe. That is my greatest concern.

Think about if you stopped a man or a woman on the street—a schoolteacher, say, or a stockbroker or a bartender—and you asked, "Will the US government come to the rescue of New York City if it were hit with a massive terror attack?"

"Of course it would," she would say.

If you asked why she believed that, the answer might be, "This is the richest and most powerful nation the world has ever seen. We have huge resources, modern technology tools, and plenty of smart people."

While all of this is true, there is an unfortunate reality, a hidden crisis, of which she is blissfully unaware. What she doesn't know is that our government has not organized those resources, technology tools, and people in such a way that it can bring them to bear in support of New York City—or any other city. This means that it cannot bring them to bear in support of her or her family or her neighborhood in the aftermath of a catastrophe.

The United States of America does not have the ability to come to the aid of its citizens in the worst instance.

It's just not there.

because this is extremistan

In this book, we have called the worst-case scenario by several names, including Maria-class disaster, catastrophe, and black swan. It has tried to convince you that you are living in Extremistan, an ultramodern society on a collision course with a range of catastrophic threats. Although you don't expect catastrophes to happen, they seem to happen all the time. Instead of preparing for them, you are relying on our government to do it. But our government is not doing it.

This is a hard message for anybody to hear, but especially for those people (like you) who are not to blame for creating the problem and who are not on the hook to fix it. Like it or not, though, you *are* on the hook. Because, someday you could find yourself in a bad spot. You could fall asleep in one world and wake up in another; inside the parallel universe, facing the dragon.

what you can do

ask the hard questions

This hard message is intended to convince you to peer out from behind your brick wall of hope, to reach out to those responsible and to hold them accountable. One way to do this is to seek them out; to find them and ask them to spend some quality time with you, walking you through, in abundant, fine-grained and multihued detail, *their plan.* You can then use the answers you get—or don't get—as a spark that ignites you to become a catalyst for change.

Reach out to the person who is in charge today: your local elected official—your mayor, county executive, alderman or alderwoman, or county judge—and (in a totally non-confrontational way) ask them for the catastrophe response plan.

> *"What is the plan to come to the aid of my family if this city is hit with a massive terror attack?"*

In all likelihood, you will get little more than a blank stare. You might then describe for them a scenario: "*Let's imagine a day, very much like today. It is a rainy Saturday afternoon in March and the Really Big One hits.*" You could say something like: "*How long will it take before help reaches my neighborhood? What is the plan to rescue people from collapsed buildings, pump out the water, get power and cellphone service back, and clear the streets?*"

If they are still with you, they might respond with something like, "*that's what [insert others] are talking about right now,*" or "*that is somebody else's [insert others] responsibility.*"

Watch out for this kind of talk, and recognize finger-pointing as pure, old-fashioned complacency. Complacency can be fixed only when there is a broad awareness of it. Condition yourself to recognize it when you see it and to call it out.

Remember our imaginary friend, Bruce? Make it your job to call out the Bruces of the world. Especially when they are the people you rely on to fix things when they go wrong.

Don't believe it when the Bruces of the world tell you that preparing for the black swan is somebody else's responsibility.

So, press on. Be determined to battle the forces of complacency as you continue with your line of questioning:

> *"Who is responsible for humanitarian relief in my neighborhood — for the seniors, individuals with disabilities, children, and families?"*

You could mention the federal ruling that requires local governments to provide to the public details of available accessible services (for the disabled and those with access and functional needs, or DAFN) to allow you to prepare your own plans.

You could even show them the US Department of Justice civil rights regulations that require equal opportunity for DAFN in all aspects of emergency planning and response.

"Do we comply with these laws? Can we provide accessible transportation during an evacuation? Where are the accessible shelters? What is the plan to canvass areas without power for disabled people who are trapped?"

To be fair, even though your local government is responsible for all these things, it doesn't have the tools it needs to do the job. You should talk with your local elected official about this. Ask them to describe the tools it has, especially its most important tool, the emergency manager.

the essential emergency manager

Every community needs a permanent emergency management team, working around the clock, to build its preparedness for disasters.

Complacency can be overcome only with continuous effort, every day and everywhere. Only focused energy, applied by disaster professionals, can create the momentum needed for true resilience. If that energy gets delayed or disrupted, the momentum is lost, and resilience evaporates.

In post-9/11 New York City, OEM blazed a pathway to resilience because it embraced the mission and because it had the resources—the people, money, and time—to do it.

The most important resource, by far, is that emergency management team. Only dedicated, full-time staff can do the things a local emergency management department must do—things like training, exercising, community and private sector outreach, field response, Watch Command, government continuity of operations, hazard mitigation; health, medical, human services, transportation and infrastructure planning, disaster mapping, information technology, logistics, and tactical teams such as urban search and rescue.

OEM as a model

Sitting at the crossroads of the world, New York City OEM attracts a continual stream of emergency managers from across the country and around the world.

Most are impressed by Henry Jackson's state-of-the-art Watch Command and EOC. But even more impressive to them are the 160 full-time professionals that make up the OEM staff. Although 160 seems like a lot, every one of those professionals owns a critical piece of the resilience mission. If the past twenty years of disasters have taught us anything, it is that no less than 160 professionals would be enough for New York City.

All those emergency managers, from across the country and around the world, envy the size of the OEM team. Few have the resources they need. Most lament their situation and wish that they had a team large enough to do for their communities the things that OEM does for New York City.

the "all disasters are local" index

Every local government—town, city, or county—needs a minimum number of dedicated full-time staff to do the things that a local emergency management department must do, such as planning, training, exercising, field response, and Watch Command.

We will use the important lessons learned in New York— lessons that can be gained only by long and varied experience with disasters—to define the minimum level of staffing for emergency management departments at the local-govern-ment level.

Based on those important lessons, for local governments in the United States, the ratio of full-time professional emer-gency managers to the resident population must, at a mini-mum, be equivalent to that of New York City's.

For the purposes of this discussion, we will call that ratio the "All Disasters Are Local," or ADAL, Index.

The ratio of full-time professional emergency managers to the resident population in New York City is twenty per million. Therefore, the ADAL Index requires the following:

Every local jurisdiction—town, city, or county—requires a minimum of twenty full-time professional emergency management staff for every million residents.

The ADAL Index is applicable to local jurisdictions with five hundred thousand or more residents. The minimum level of staffing for emergency management departments for local jurisdictions (cities and counties) with less than five hundred thousand residents is seven (a director, deputy director, planning chief, training chief, exercise chief, mitigation and recovery chief, and administrative/grants chief). The team-of-seven requirement assumes that the emergency management department can rely on other government agencies for critical missions such as field response, disaster mapping, logistics, and information technology.

how does your city rate?

When comparing existing staffing levels to the ADAL Index, you'll find that nearly every local emergency management department in the United States is understaffed. The Philadelphia Office of Emergency Management, for instance, needs thirty-one full-time professionals to be fully staffed, the Phoenix OEM needs thirty, and the Dallas OEM needs twenty-five.

The ADAL applies to counties too. The Los Angeles County Office of Emergency Management needs two hundred full-time professionals to be fully staffed, the Cook County (Illinois) Department of Homeland Security and Emergency Management needs 105 staff, the Harris County (Texas) Office of Homeland Security and Emergency Management needs eighty-seven, and the Miami-Dade (Florida) Office of Emergency Management needs fifty-two.

find out if your local emergency managers have what they need

Calculate the ADAL Index for your city or county. It's simple: take the number of residents of your city or county and divide by one million. Multiply that number by twenty.

Compare your result to the number of full-time staff in your local emergency management department.

How did your city or county rate? Scores of fifteen or higher signal that your local elected officials have made a commitment and your jurisdiction is on the pathway to resilience. If your community's score is five, for instance, you have less than half of the resources you need.

Come to the meeting with your local elected official armed with this important information. Tell them, "Our local emergency managers do not have what they need."

Point to the example of the New York City OEM and ask them, "What if…? What if we were hit with a Superstorm Sandy? Or something else, something much worse?"

what if…the mother of all disasters?

On January 13, 2018, an emergency alert sent to every cell-phone in the state of Hawaii turned a serene Saturday morning in paradise upside down.

> "Emergency Alert. BALLISTIC MISSILE THREAT INBOUND TO HAWAII…SEEK IMMEDIATE SHELTER…THIS IS NOT A DRILL…"

Somebody pushed the wrong button and, at 8:07 a.m. local time, an automated message informed people across the densely-populated island archipelago that everything they knew and loved was about to be incinerated by forces beyond their control.

In the aftermath of the incident Vern Miyagi, Administrator of the Hawaii Emergency Management Agency, stood beside his boss, Governor David Ige, in front of a crush of television cameras: "The warning was a mistake," Miyagi said.

If there is a silver lining to this mistake, it is that it forced Hawaiians to peek over the top of their brick walls. For nearly thirty-eight minutes, they were forced to engage this very real threat and ask themselves, *"What do I do now?"*

know your nuclear first steps

Those of us who are not fortunate enough to live in Hawaii should also think about peering out from behind our brick walls. The possibility of a nuclear attack is real. It hasn't happened, but it will. Thank goodness it didn't happen on that serene Saturday morning in Hawaii, but chances are that it will happen somewhere and someday. Do you know what you would do if you got that text?

Disaster professionals such as myself, prepare for disasters by putting ourselves in the middle of an imagined disaster so that we start to figure it out ahead of time, instead of in the fog of war. We call these scenarios. Scenarios can be useful for you too.

One of the best things that you can do, right now, is to take a few minutes to work the same process with that worst-case scenario:

"It's 2:35 p.m. on a rainy Saturday afternoon in March. You just sat down on the couch, and are seriously contemplating a nap, when your cellphone goes off. But instead of the familiar ring tone, it is sounding that very loud Emergency Alert System blare. You stare at the message on the screen:

'Emergency Alert. BALLISTIC MISSILE THREAT INBOUND... SEEK IMMEDIATE SHELTER...THIS IS NOT A DRILL...'"

Okay, now what? If the unthinkable happens and an emergency alert about an incoming missile turns your world upside down, what will you do?

First, you should know that you will survive; if you take some immediate steps to protect yourself. In that moment, there are three things that can protect you: time, distance, and shielding. Because there is no time to run, you can minimize your exposure to the radiation with shielding. Get thick walls of concrete and brick between you and the outdoors as quickly as possible. Better yet, go to the nearest basement or to an interior room with no windows. Then stay where you are and listen to news reports and local officials about when it is safe to leave. Remember the mantra: *get inside, stay inside, and stay tuned.*

When you find yourself in that parallel universe, remember that doing something is always better than doing nothing. Once you are out of immediate danger, your best first step is to pause. Take a deep breath, slow your breathing, and begin to walk back some of the damage done by your lizard brain. Tell yourself that you are going to use that highly developed brain of yours—in combination with your senses—to get your life back.

In the event of an actual nuclear strike, residents would be instructed to go inside and remain sheltered for up to fourteen days or until they are told it is safe to leave.

Fourteen days is a long time.

There are some simple things that, if done beforehand, could make those fourteen days go a lot easier. If you take a moment to ponder what it would be like to be in that situation, eventually some of these things will start to take shape in your mind. Things you would have to do, the people you would have to connect with, and the stuff you would need. Here's yet another mantra for you: *Know what to have in your hand, what to have in your head, and what to have in your home.*

what you should have in your hand in case of emergency: a go bag

A "Go Bag" is a collection of essential items you want with you if you had to leave your home or workplace in a hurry. Your Go Bag should be sturdy and easy to carry, like a backpack or a small suitcase on wheels. You'll want to customize the contents of your Go Bag for your needs, but some of the important things to consider include copies of your important documents (insurance cards, birth certificates, photo IDs, proof of address, and the like), an extra set of car and house keys, photocopies of credit cards, a couple of hundred dollars in cash (in small bills), bottled water, nonperishable food (such as energy or granola bars), a flashlight, a battery-operated AM/FM radio, a list of the medications each member of your household takes, and a first-aid kit.

Individuals with disabilities should consider items such as food and drink for special dietary needs, medical bracelets or identifying alert items, extra eyeglasses, and batteries for hearing devices.

Don't forget to stock up on essential items your pet may need, such as special, an extra collar and leash, medications, a crate or sturdy carrier, a blanket, and a recent photo of your pet (to find him or her if you are separated).

what you should have in your head in case of emergency:
a family emergency plan

A family emergency plan is a communication plan for your family. Creating your family emergency plan starts with one simple question: "What if?"

> *"What if something happens and I'm not with my family?" | "How will I reach them?" | "How will I know they are safe?" | "How can I let them know I'm okay?"*

During a disaster, you will need to send and receive information to and from your family. Your family emergency plan should include a plan to get out of the house quickly and safely, the name of a place to meet if your family is split up, a phone contact list with contact information for one or two out-of-town relatives, and a plan to communicate via social media platforms.

Once you have created a plan, convince your loved ones to sit down and talk through it, and keep it updated.

what you should have in your home in case of emergency:
an emergency supply kit

An emergency supply kit is a collection of basic items (including food, water and medications) that your family would need in the event of an emergency. It could take days for local officials and relief workers to reach you, and you will need to be able to survive on your own.

Assemble your kit now, since you will probably not have time to shop after that annoying Emergency Alert System buzz comes over your cellphone.

The ideal emergency supply kit would contain everything you would need to manage for a couple of weeks, but it will probably be depleted within a few days, after which you will need help.

beyond personal preparedness:
community preparedness basics

In a post-catastrophe environment, the first help that comes to your home will almost always be from your neighbors. That is why community preparedness matters. These days, as we have said, the Whole Community—including individuals and families, businesses, faith-based and community organizations, nonprofit groups, and schools—have a lot on their plates. These full plates—along with things like complacency and good old-fashioned procrastination—thwart real progress in preparedness.

The missing piece, of course, is leadership. To the extent that community preparedness is happening, it is because of the individuals who give their time and talent to spearhead grassroots efforts.

Imagine these impassioned individuals convening groups of people in living rooms and church basements around the country. They are pushing past the inertia and the passive resistance with detailed descriptions of the dangers that lie beyond the brick wall. They are empowering their communities through the same building process that OEM used to lead city agencies through in post-9/11 New York City: planning, practicing, gathering resources, and creating real capability.

If this seems like mad work, that is because it is. Who has the time or the energy these days? As difficult as it may seem, reaching out to your neighbors to plan for an emergency doesn't have to be that hard. According to Mitch Stripling, Assistant Commissioner at the New York City Department of Health and Mental Hygiene, instead of going door-to-door to rally your neighbors, all you really have to do is get in front of some of the groups that are already active in your neighborhood, like community boards and PTAs, churches, or even book groups or sports teams. The key is to get preparedness on the agenda in a fun and user-friendly way.

> *"You can schedule a disaster-focused P.T.A. meeting, or a bar trivia night about disasters. Your local coffee shop could host a Godzilla Awareness Party."*[179]
>
> —"How to Prepare Your Community for a Disaster,"
> *The New York Times,* February 15, 2018

Once you have their attention, you can get them to start thinking about a community preparedness plan. You don't have to be specific, the goal is for everyone to get a better understanding of their neighborhood and how they can help each another before, during, and after a disaster.

According to Stripling, a good community preparedness plan includes a contact list so that everyone knows how they will connect, using text messages, landline phones, and rendezvous points.

Keep in mind that disaster impacts are worse for people who are already at risk, your elderly neighbors, people who live alone, the disabled and those with access and functional needs.

Make it a point to forge an ongoing connection with your at-risk neighbors before the disaster strikes. Get to know them and their needs. Check in on them in the early hours of a disaster. By reaching out to high-needs populations, you are on the front lines of the response. You can assist directly or identify at-risk individuals and families for the door-to-door teams of relief workers that will be canvassing your neighborhood.[180]

other things that you can do

You play a critically important role in preparedness for yourself and your family. If you have the time and inclination, here are some more ways you could improve preparedness for your city, your state, and your nation.

support real solutions to the DAFN crisis

Where possible, work to increase awareness of the DAFN crisis. By understanding the nature and magnitude of the challenge, and the struggle that the people who are responsible for doing these things are dealing with, you can begin a process of awareness that is the prerequisite for a national solution. Support proposed solutions such as a formal and funded national Red Cross–government partnership.

discover the joy that comes from service to others

Consider volunteering with one (or more) of the many organizations that care for the poor and the underprivileged, seniors, individuals with disabilities, children, and families in disasters.

Although some will say that service to others is an obligation, whether it be moral, spiritual, or social, I take the opposite view. My time at the Red Cross showed me that to serve others is a privilege. He or she who serves always benefits more than he or she who is served.[181]

One of the most rewarding experiences in this life is to enter the disaster zone to connect with affected individuals and families. Support the organizations that do that. In addition to the American Red Cross, Americares, Feed the Children, and Team Rubicon, you should consider faith-based organizations such as the Salvation Army, Tzu Chi, Lutheran Disaster Response, Catholic Charities, Samaritan's Purse, and the Southern Baptist Convention, or community-based government organizations such as Community Emergency Response Team and the Medical Reserve Corps. Contact your local Voluntary Organizations Active in Disaster, or VOAD, for information about groups that are active in your area.

the hardest job in the world

Scientists tell us that two million years ago, *Homo erectus*, our ancient ancestors, first walked upright upon the land. By that time, the brick wall of hope was already a well-established fixture within the mind of every *Homo erectian*—disasters having been a part of everyone's life for as long as anyone could remember.

Humanity has been responding to disasters since that time, and our track record is mixed at best. Our track record with worst-case disasters, however, is not mixed. Throughout our long history and to the present day, catastrophic response is an unbroken series of abject failures.

No generation has ever effectively assisted massive numbers of its fellow human beings in the midst of catastrophe. There are some very good reasons for this. Catastrophes bring with them unique challenges. They are different from other disasters because they affect everyone at the same time; they cross political boundaries and they create a demand for resources that greatly exceeds what is immediately available.

That is why catastrophic response is the most difficult of human endeavors. It requires that disaster professionals find and fix a multitude of urgent problems and unmet needs across a vast parallel universe…

…where the normal rules of logic don't apply

…where need is a hundred times greater than the resources at hand

…where a hundred times more problems exist than anyone has the capacity to engage

…where time is elastic, slowing down, then suddenly flying by

…where cellphones don't work

…where roads are blocked…

…and the fact that it's difficult is no excuse for not doing it.

Despite the challenges, we're duty-bound to assist our fellow human beings in their time of greatest need, and we must do it with dignity and respect.

So…what are we waiting for? Let's get to work.

Disaster professonals need to stop spending so much time trying to predict the future, searching for bugs in the software or targets that need hardening, and spend more time learning how to reconfigure themselves to confront the unknown in a complex environment.[182]

There aren't two ways to do this; there is only one. Disaster professionals all over the country must come together into a massive team of teams, an incident organization, a Great Machine; the bigger the catastrophe, the bigger the machine. The United States of America needs the ability to bring together a Great Machine—the size of a Google or even a Walmart—with the ability to communicate up, down, and across the organization in one day.

All our assets must be engaged: first responders, government agency staff, National Guard soldiers, aid workers, construction workers, private employees, and volunteers from every city, town, county, borough, and parish across the country. The challenge is that every government agency, nonprofit organization, military battalion, and private company is occupied by and absorbed in a daily mission. Every piece is a separate silo, and only a compelling need can draw it away from that daily mission.

By coming together now, disaster professionals can create muscle memory around this process—a straightforward process clearly understood by all—to bring the resources of the nation to bear in the early hours of the disaster, to reach deep into the system to buy, beg, and borrow everything we will need.

somebody needs to do for the nation
what OEM did for New York

If all of this sounds daunting, it is. Yet, who can deny the nature of the threat or the urgency of the need? Again, there aren't two ways to do this; there is only one.

Which means this sort of comprehensive, proactive, integrated, and all-of-nation planning is going to happen. It's just a question of when—and whether it will be done the easy way, by figuring it out beforehand, or the hard way, in the aftermath of the next black swan.

We can do this.

The United States of America can build a Great Machine that can bring together a massive team, get on the ground, and knock on every door.

We can connect to leadership and break down silo walls and, by doing so, reconnect our citizens and reclaim an orderly society. To do it, though, we need a new program; a national program, as big and ambitious and all-encompassing as Apollo...

"We choose to go to the moon in this decade and do the other things, not because they are easy, but because they are hard, because that goal will serve to organize and measure the best of our energies and skills, because that challenge is one that we are willing to accept, one we are unwilling to postpone, and one we intend to win."[183]

—John F. Kennedy, September 12, 1962

It will take commitment—nothing less than the commitment we saw from President Kennedy and the nation in September 1962. This new Apollo Program would build a true national disaster response system, a 21st-century technology-led Great Machine that can get the right help to the right place at the right time to our fellow human beings trapped in every type of disaster.

epilogue

the polaris catastrophe reimagined

OEM sounds the alarm | *4 p.m.* | *Thursday, August 29* | *Watch Command* | *OEM headquarters* | *Brooklyn, New York City*

An ultrahigh-tech nerve center akin to NASA Mission Control, New York City's Watch Command monitors police and fire broadcasts, media, social media, 911 systems, and weather forecasts. It is the eyes and ears of the city. Its one and only job is to alert everybody—city, state, and federal officials, and the public—to real or potential threats.

Taina Galarza, a thirty-something Brooklyn native and OEM veteran, was promoted to the rank of Watch Command Supervisor three years ago. She and her team are an hour into Tour 3, the 3 p.m. to 11 p.m. shift, at OEM headquarters in Brooklyn, when her control panel lights up and a cacophony of alarms fills the room. She stares at the message on her screen:

"BALLISTIC MISSILE THREAT INBOUND TO NEW YORK CITY... SEEK IMMEDIATE SHELTER... THIS IS NOT A DRILL..."

Damn, she thinks. *Not another false alarm. NORAD*[184] *told us they had fixed this. I don't need this right now.*

Her thoughts are interrupted by a violent earthquake-like tremor, followed by the roar of a rumbling explosion. The floor in Watch Command jumps and shakes. The screens in the windowless room go suddenly white as an intense flash of light passes over the cameras.

Overhead lights flicker as the emergency power generators kick on. Computers that weren't supposed to go down try to reboot; but most of the screens stay black or are filled with gray snow.

Galarza glances at a remote camera mounted on the control tower at John F. Kennedy International Airport and pointed at Midtown. The screen is filled with a towering column of fire.

A wave of fear washes over her. She talks aloud to keep the panic at bay, "Heads up people...Camera 12 showing what looks to be a nuke...direct impact on Midtown."

She looks around the room at her team, two Watch Commanders and a public warning specialist. One is a veteran emergency medical technician with years on the job; the others are young, just getting their start in the disaster business. All are scared.

"This is it guys—the job we've been training for. We have to go to work now. Let's do this," Galarza says.

Three minutes after the blast, Watch Command issues its first warning.

"Everyone—individuals, families, residents, visitors—get inside, stay inside, and stay tuned."

"First responders—including police and fire personnel—get inside and await further instructions."

president is handed a plan | *4:24 p.m.* | *Thursday, August 29* |
Willard Hotel | Pennsylvania Avenue NW, Washington, D.C.

The aide-de-camp walks slowly across the stage to the podium and whispers the code word "Trinity" into the president's ear as he is giving a speech to a packed auditorium in downtown Washington, D.C.

Without missing a beat, the president immediately concludes his speech. He waves and smiles broadly as he walks briskly off the stage.

His mind races as aides show him a series of cell phone pictures that were sent by the Pentagon. They show satellite imagery of a huge fireball and starburst-patterned blast wave radiating across the width of Midtown Manhattan.

The president and his team have drilled for this moment a handful of times, and he knows what it means. "Somebody please tell me that this is a drill."

"I wish we could say that it is, sir. Those pictures were taken by our surveillance satellites within the last fifteen minutes. They are showing what appears to be a nuclear explosion in New York City. Based on those images, we estimate the yield of the device to between five and seven kilotons."

After a long pause, the presidents asks, "Who did it?"

"No confirmation as of yet, Mr. President. We have a couple of ideas tho."

With the worst-case scenario clearly upon him, the president is momentarily speechless. Finally he manages, "What are doing about it? Like, right now?"

"We have activated the Nuclear Response Plan, sir. We are getting everybody and everything we have all-in the response. Here is the first situation report...."

oem finds the plume | *4:26 p.m.* | *Thursday, August 29* | *Watch Command* | *OEM headquarters* | *Brooklyn, New York City*

Nuclear First Steps is the core drill in the emergency management lexicon, and OEM had conducted it repeatedly over the years. So, as OEM staffers continue to issue an immediate action message, first in English and then in multiple other languages (this is step one: sound the alarm), they know that step two is even more critical.

Where is the plume? They know that the column of fire has propelled debris from the blast high into the upper atmosphere. Over the next several hours, fine grains of sand-like particles will rain from the sky as highly radioactive fallout. They need to predict the location of the fallout plume and tell people in those areas to go inside. Anyone caught outside in the plume, also known as the dangerous fallout zone, or DFZ, will be exposed to a lethal dose of radiation.

Weather radar shows sustained winds out of the northwest at five to ten miles per hour. Based on this, Watch Command predicts[185] the plume is moving from Manhattan across the East River and over southern Queens and all of Brooklyn.

Although damage from the electromagnetic pulse had disrupted its communications and many of its systems are down, Watch Command uses every available means[186] to transmit the plume warning to people across Brooklyn, Queens, and Long Island. It also sends the message to its counterparts in New Jersey, Connecticut, and at the New York state Warning Point in Albany. They all blast it via the national Emergency Alert System, public warning systems (for example, NotifyNYC, NY-Alert, and NJAlert), broadcast voice mail, email, emergency text message alerts, and social media (including Twitter and Facebook).

The plume warning instructs who should evacuate and in what direction, the protective measures for first responders, and where it is safe for first responders to operate.

Within an hour, they attach a map to the message. The map, labeled "NYC NRP 082921#1," is a compilation of everything they know or have predicted. It shows the blast location surrounded by concentric circles that define what it called Light Damage (LD), Moderate Damage (MD), and No-Go Zones (NGZ). The map also shows the DFZ extending dozens of miles downwind over the densely populated areas of Brooklyn, Queens, and Long Island. NYC NRP 082921 will be updated and blasted out several times a day. It will ultimately become one of the most important documents in the history of 21st-century America.

Many years later, Taina Galarza would be awarded the Congressional Gold Medal—the highest civilian award of the United States—for her actions in Watch Command on August 29. The scientists and historians who would write the history of the Polaris Attack would agree that the actions of Galarza and the rest of the Watch Command team, along with similar teams in state warning points, regional intelligence and operations centers, and the FEMA watch center, saved hundreds of thousands of human lives that day.

state and fema step up | *August 29–30* | *Emergency operations center* | *OEM headquarters* | *Brooklyn, New York City*

Watch Command spent a lot of time on that hot August afternoon trying to reach the Duty Team chief. On standby around the clock, the Duty Team is made up of OEM emergency managers who coordinate the city's response from the New York City EOC.

As Deputy Commissioner, Henry Jackson was responsible to lead the White Team. The White Team was "up" and therefore owned the job. Jackson's job was to turn everything on and get everyone "all-in" the job. To do this, he needed to turn on the Great Machine and plug everyone into it.

As the core of the Great Machine, the New York City EOC is the White Team's tool to get the right information to everybody who needs it—the right stuff in the right place at the right time—and to find and solve problems.

For this incident, it's hard to overstate the difficulty of even one of these things, let alone all three.

The White Team starts by clicking into place, turning on the Great Machine, and activating the New York City Nuclear Response Plan, or NYC NRP. The NYC NRP starts by connecting to every available local, state, and federal agency, and every private, nonprofit, nongovernmental, faith-based, and volunteer organization. All of these have designated responsibilities within the NRP.

But, as that day wore on, the White Team did not do any of these things.

That is because Watch Command continued without success to try to reach Henry Jackson and the rest of his White Team.

Fortunately for New York, others owned the job, too. So as specified in the National Nuclear Response Plan, a combined FEMA/ New York State team took the reins and opened a regional joint field office (aka the "Mega JFO"), a massive EOC, at Westchester Community College in Valhalla, New York, twenty-five miles north of Times Square. Representatives from all over the region and the nation began to report there and, within 48 hours, it housed over four thousand disaster professionals.

As per the plan, the Mega JFO activated local and state stockpiles and supply stand-by contracts. It called for every asset available through the Emergency Management Assistance Compact (EMAC) and local and regional mutual aid agreements. It activated the strategic national stockpile and strike teams, Task Forces, National Guard, Army, Navy, Air Force, and Marines.

It activated forward operating bases in strategic locations around New York City to accept and organize a massive influx of resources—personnel, materials, supplies, equipment, and vehicles—and to integrate them into the response.

It also enabled a unique postnuclear incident command by convening elected officials in a leadership cell composed of decision-makers from the federal government, affected state and local jurisdictions, and representative elected officials. These officials met initially via conference call and then moved to in-person meetings at the Mega JFO.

The Great Machine, centered at the Mega JFO in Valhalla, New York, facilitated an integrated, all-of-nation response.

the president takes the reins | *12:49 a.m.* | *Friday, August 30*
| *Secure Situation Room* | *White House, Washington, D.C.*

The president is back in the Situation Room deep below the White House. In a conference room packed with secretaries and generals, Maryanne Tierney, the FEMA Administrator, is talking about the National Nuclear Response Plan. She is describing how the local government did its job. It sounded the alarm, recognizing the incident and getting everybody inside. It found the plume on a map and determined protective actions. And it turned everything on.

"What are *we* doing?" the president asks.

"We are leading the response, Mr. President," Tierney replies.

"Thank god," says the president. "I was worried that you were going to try to tell me that we were waiting for the state or some other nonsense. I want everybody we have and everything we got in New York and in the fight. I want our best people working on this, our best thinking. I don't care what anyone thinks they should be doing right now. Getting our arms around this is our one and only priority right now."

The president looks around, making eye contact with everyone at the table. "No foot dragging, no excuses, no bull-shit. Are we clear about that?"

Heads nod vigorously.

"We have been practicing this for over a decade," Tierney explains. "All agencies and departments, every federal asset, is either being made available or is en route to New York." Tierney goes on to describe the plan and tell the story of the Great Machine.

"I can't tell you how relieved I am by this. This is exactly what we need to be doing, exactly what the American people expect of us."

"Yes, sir. But there is a long, hard road ahead," Tierney said.

"Indeed, the longest and hardest road imaginable. But the important thing is that we are on it and we know where it is going," said the president.

The Secretary of Defense adds, "This is a very dangerous situation, Mr. President. Radiation levels are off the charts. Isn't that right, Juliana?" He looks at the EPA administrator.

"We need to be mindful of the risks, Mr. President," the EPA Administrator says. "There are hazards in and around the blast site, and we need to protect our people."

"Absolutely we need to protect our people. You're telling me we've been working on this for ten years and haven't yet figured out how to protect our people?"

"It's not that we don't know how; it's just that we have no experience with these kinds of numbers," says the secretary.

"Nobody does," Tierney says, "but our Safety Task Force has a laser focus on this, and a Health and Safety Program is already in place. This is just one of dozens, even hundreds, of issues that need to be resolved. That's what the Mega JFO is doing right now. Our best course is to continue to get our best people and our best thinking plugged into the teams assembled there. That's the plan."

"Okay. This is no time to try to figure out a new game plan. If we've got a plan, then we need to stick to it," the president says.

"We do and we are sir," Tierney replies.

"Good," said the president, "Now, where is this Mega JFO?"

"Twenty-five miles north of Ground Zero, in Valhalla, New York."

He turns to his aides.

"Let's go."

mega JFO leads the response | *Noon* | *Friday, August 30* | *Mega JFO* | *Westchester Community College* | *Valhalla, New York*

In the early days, the Great Machine, centered at the Mega JFO, continued to work the plan.

It connected to command posts around the perimeter of the affected area. It established new ones where needed to support field operations.

It supported search-and-rescue operations as they began to extricate victims trapped in collapsed buildings.

It facilitated spontaneous evacuations by coordinating crowd control, messaging, and exit points.

It coordinated with the National Guard and the US Army personnel to create a security perimeter and conduct traffic and crowd control.

It continued to transmit messages to the public about evacuation and mandatory shelter-in-place orders.

It began radiological data collection based on the map, NYC NRP 082921 and determined the protective equipment needed for response personnel based on the hazards on the ground in the affected areas.

It connected to the medical community and established triage locations to provide patient triage and treatment.

It coordinated with hospitals outside the affected area, and across the nation, to manage the surge of sick and injured, evacuating large numbers of patients to hospitals outside the impacted area.

Given the spectrum of injuries (including blast, radiation, and thermal injuries), the Mega JFO coordinated a common approach to medical standards of care.

It connected to forward operating bases around the city to accept and organize the massive influx of personnel, materials, supplies, equipment, and vehicles and to integrate them into the operation.

the president speaks to the nation | *Noon* | *Friday, August 30* | *Hankin Arts Center* | *Westchester Community College* | *Valhalla, New York*

The spacious theater at the Mega JFO is standing-room-only as the president walks across the stage to the podium.

"I just received an update from Homeland Security Secretary, Secretary Ojami and the other Cabinet secretaries involved on the latest developments in New York City.

"I can tell you that our thoughts and prayers go out to the people of New York City and others affected by this enormous tragedy. There is so much that we don't yet know, but here is what we know now. Yesterday afternoon at approximately 4:04 p.m. local time, a nuclear device with an explosive yield of approximately ten kilotons, equivalent to ten thousand tons of TNT, detonated in New York City's Times Square. We don't yet know who the perpetrators are, but the effects of this evil attack were immediate and devastating. Most of midtown Manhattan is damaged or destroyed. Thousands, possibly hundreds of thousands, have been killed and many more have been injured and need urgent medical care. Response efforts are hampered, however. The city is cut off, with all roads, bridges, and tunnels closed.

"I have spoken with the governors of the states of New Jersey and Connecticut, and we are attempting to contact elected officials both in the city and in New York state. We have been unsuccessful so far, but our efforts are continuing.

"I want to recognize the heroic efforts of the New York City Office of Emergency Management in sounding the alarm and getting everybody inside. Its instructions came immediately after the blast and were instrumental in saving lives and keeping people safe. The hazards from the blast continue to threaten us, and I am calling on all residents of the affected areas to continue to heed those critically important directives from local authorities. If you are told to shelter in place, you should continue to do so. If you are told to leave, you should evacuate as soon as possible. Our scientists are working closely with those local authorities to get you where you can be safe as soon as it is safe to do so.

"Now, I want to talk to you today about what we are doing, as a nation, to respond. Yesterday afternoon a few minutes after the attack, I activated the National Nuclear Response Plan. I assigned former general Stanley McChrystal as the National Incident Commander, and the federal government is now leading the response to this catastrophe. That effort—a coordinated all-of-nation response—is headquartered here, in Valhalla, New York, twenty-five miles due north of the site of yesterday's attack.

"Our efforts are now focused on three priorities. First and foremost we are saving lives. We are working to extinguish the fires in the areas where it is safe to do so. In addition, we are conducting search-and-rescue operations in those areas."

Katherine Holders of ABC News calls out, "The images coming out of New York City are horrifying, Mr. President. Thousands of people are dying in the streets. What are you doing to help them?"

"Right now, search-and-rescue teams are extricating victims trapped in collapsed buildings and under debris. We are establishing triage locations and providing patient treatment for the injured. We are organizing hospitals outside the affected area, and across the nation, to manage the sick and injured, and we're working to evacuate large numbers of patients to hospitals outside of the impacted area."

Holders follows up, "What about the people who walked through the Lincoln Tunnel to Weehawken? The police there are not letting them stop. Where are the shelters for them?"

"We are working to support the evacuation of any remaining citizens from the affected area," the president replies. "I want to thank the states of New Jersey and Connecticut for agreeing to provide shelter for some of those citizens. I call on other surrounding states, particularly Pennsylvania and Massachusetts, to do more in this moment of national crisis.

"We are moving toward providing emergency assistance to these individuals and families. We are focused on helping them and reunifying them with their families.

"We're just beginning to organize a massive decontamination operation to screen and decontaminate contaminated persons.

"We have activated our National Mass Fatality Plan to recover decedents and return them to their next of kin.

"We are beginning to assess damage to our critical infrastructure, including roads, bridges, subways, and airports, hospitals, communication systems, electric power and water systems, and of course, to our buildings."

Finally, Holders blurts out, "When will this situation be under control?"

"What I can tell you now is that this recovery will take a long time. This recovery will take years. But we have also been planning for this for many years in partnership with states and locals and we were prepared. All agencies and departments—every federal, state, and local asset—are either being made available or are en route to New York.

"I want to assure you that we have the very best people from across this great nation working here, in the regional joint field office, with us. Our best minds and our best thinking are dedicated to this. Everybody and everything we've got. Helping the people of New York is the sole focus of the United States of America right now."

The president pauses and looks around, making eye contact with the crowd.

"I will take any further questions at this time…"

mega JFO works the plan | *August to September* | *Mega JFO* | *Westchester Community College* | *Valhalla, New York*

As those early hours stretched into days, the Great Machine continued to work the plan. It began a massive debris management operation to remove contaminated debris from blocked streets to facilitate search and rescue and medical evacuations. It organized and managed the evacuation by identifying the neighborhoods and the communities that needed to move. It managed a vast humanitarian relief operation to shelter, feed, and distribute relief supplies to surviving populations. Finally it organized the process of disseminating information about the victims to surviving family members.

the situation starts to improve | *August to September* | *Areas around New York City*

Three weeks pass, and the situation starts to improve. New Jersey and Connecticut have been overrun with people, but the Delaware, Pennsylvania, Rhode Island, and Massachusetts borders are open, and the states are starting to absorb them.

Fewer people are sleeping on the ground and under tarps. Food and water and medical supplies are plentiful, and more shelters are opening in the surrounding towns.

Meanwhile, in and around Midtown Manhattan, individual and team stories of heroism and sacrifice are emerging from the horrific situation. First from FDNY and NYPD, and then from outside teams who came in before the borders were closed.

With the US government leading the charge, surrounding states are stepping up and assisting wherever they can.

The US Military's Joint Task Force Civil Support is playing a huge role. As it pulls in assets and resources from around the country and the world, it is beginning to execute a series of clear missions that support all the activities described here. Originally it confined its efforts to the periphery for fear of putting its soldiers into harm's way, but NYC NRP 082921 has allowed it to move into areas it knows are safe. At the same time, the Mega JFO is conducting around-the-clock training to protect people who must work in contaminated areas.

The Mega JFO in Valhalla becomes a joint information center that provides clear and consistent communication to the public through all available information channels in a variety of formats and languages.

It works across the Great Machine to provide consistent answers to the hundreds of questions from the media and the public.

the nation triumphs | *September to October* | *Mega JFO* | *Westchester Community College* | *Valhalla, New York*

Something about the battle rhythm of the Mega JFO appeals to the president. He connects to the pace and the process and spends more and more time there, taking a hands-on role in the response. He even ventures into New York City at one point, donning protective equipment and entering the No-Go Zone.

As the weeks pass, the world watches as the situation improves. Politicians in unaffected states begin to call for increasing their involvement. The governors compete to show who is doing more for New York City and the nation in this time of crisis.

As the weeks stretch into months, one thing is clear. The mighty USA, though shaken, will emerge even stronger.

Significant scientific breakthroughs emerge from the great minds at the Mega JFO. The president and the US lead the way to a nuclear nonproliferation treaty and, eventually, to a nuclear-free world.

afterword

everyone needs to know the nuclear first steps

For nearly six years, I was the New York City representative for a groundbreaking initiative with more than thirty counties and states, called the New York-New Jersey-Connecticut-Pennsylvania Regional Catastrophic Planning Team, or RCPT.

The RCPT did a lot of thinking about this problem, and a lot of building. We designed and built the tools and the processes to bring together the federal government, along with multiple states and the locals, into a coordinated disaster response. We built plans for critical operations, such as debris management, disaster housing, sheltering, evacuation, infrastructure protection, and mass-fatality management.

And we built the NRP, the Nuclear Response Plan.

The NRP contains a structure, guidance, and dozens of tools. It leverages recent scientific breakthroughs that tell us what we should do in the early minutes of an attack to save thousands of lives.

This is a national plan, but it has not been put into practice. It has not been absorbed and used as a basis for a national program by which all states and the federal government come together at the same time to practice. To clearly understand those straightforward Nuclear First Steps:

First, *sound the alarm; recognize the
  incident and get everybody inside.*
Second, *find the plume and tell everybody where it is.*
Third, *activate all plans and get everybody all-in.*
Fourth, *conduct a massive field response.*
Fifth, *manage the surge in information,
  resource needs, and problems.*
Sixth, *convene all levels of government.*
Seventh, *get help from every other state.*

Every disaster professional should understand and be able to execute these first steps. All players at every level of government and beyond must be able to immediately activate, organize, and flawlessly execute them. To do that, they need to practice *now* exactly what they are going to do *then*.

doing something is always better than doing nothing

In the case of a catastrophic response, doing something means bringing the nation together in our most desperate moments to save vast numbers of our fellow citizens.

Catastrophic response is the most difficult of human endeavors, an instantaneous and complex marriage of strategy and tactics. Within minutes, an incident organization (a Great Machine) must achieve synchronicity with elected leadership, across a breathtaking array of crises and a huge affected area.

The nature of the catastrophe is that it will start suddenly and with great intensity, with the biggest problems and greatest needs coming in its earliest minutes and hours.

These early minutes and hours—the so-called golden hours—will be a time of maximum chaos. The actions we take in the golden hours will determine our fate.

endnotes

---

1   Steven Starr, Lynn Eden, Theodore A. Postol, "What would happen if an 800-kiloton nuclear warhead detonated above midtown Manhattan?" *Bulletin of the Atomic Scientists*, 25 April 2015, accessed at http://thebulletin.org/what-would-happen-if-800-kiloton-nuclear-warhead-detonated-above-midtown-manhattan8023

2   Located two miles due south of Times Square, 18 Squad is a specialized unit that can not only fight fires but also rescue people and work in hazardous conditions like those it was about to face.

3   Adam Stone, "Catastrophic Power Outages Pose Significant Recovery Challenges," *Emergency Management*, January 21, 2013

4   United Nations, Department of Economic and Social Affairs, Population Division, "Risks of Exposure and Vulnerability to Natural Disasters at the City Level: A Global Overview," Technical Paper No. 2015/2, 2015, accessed at https://esa.un.org/unpd/wup/Publications/Files/WUP2014-TechnicalPaper-NaturalDisaster.pdf

5   Average Weekday Ridership, MTA | Facts and Figures, Subway (5,655,755/day), bus (2,445,320/day), LIRR, Metro-North, NJT railroads (266.8 million/year), accessed at http://web.mta.info/nyct/facts/ridership/

6   Ashley Feinberg, "Seamless Is Down and People Are Freaking Out," *Gawker*, 29 August 2015, accessed at http://gawker.com/seamless-is-down-and-people-are-freaking-out-1727508643

7   An example of the latter is the change in New Yorkers' perceptions of hurricane risk since Hurricane Sandy made landfall in October 2012.

8   Seamless does not prepare or deliver food. Once a user submits an order, it is automatically sent to a dedicated computer terminal at the restaurant. The restaurant confirms the order with Seamless and then prepares and delivers the order.

9   "Is Your Manufacturing Supply Chain Vulnerable to Disruption?" as Appeared on *Bloomberg Business* as Part of the New Perspectives on Risk Series, 2018, The Travelers Indemnity Company, accessed at https://www.travelers.com/resources/business-industries/manufacturing/manufacturing-supply-chain-vulnerabilities: "This sector deals with many components and sources in its supply chain, and a disruption to any single piece could derail the whole process—and the daily life of millions of people along with it."

10 Patricia Hoffman and William Bryan, Large Power Transformers and the U.S. Electric Grid, Infrastructure Security and Energy Restoration, Office of Electricity Delivery and Energy Reliability, U.S. Department of Energy, June 2012, accessed at https://energy.gov/

sites/prod/files/Large%20Power%20Transformer%20
Study%20-%20June%202012_0.pdf

[11] Jay Apt, "Blackouts Are a Fact of Life. Let's Deal with Them," *The Wall Street Journal*, 17 September 2012, accessed at http://online.wsj.com/article/SB100008723 96390444860104577560951695371724.html

[12] Ted Koppel, *Lights Out: A Cyberattack, a Nation Unprepared, Surviving the Aftermath*. New York: Random House, 2015

[13] Hugh Byrd and Steve Matthewman, "Energy and the City: The Technology and Sociology of Power (Failure), *Journal of Urban Technology* 21, no. 3 (2014): 85-102, 4 November 2014: "Reliability and profits are at cross-purposes—single corporations put their own interests ahead of the shared grid, and spare capacity is reduced in the name of cost saving. There is broad consensus among industry experts that this has exacerbated blackout risk."

[14] Byrd and Matthewman, webpage: "Electrical power blackouts are often reported as human errors or as technological shortcomings, so the problem is either reduced to the level of individuals, or to nuts and bolts. This binary of blame—people or hardware—only obscures the systemic nature of network failures, which are the outcome of relations between people, technical systems, resources, institutions, regulatory frameworks, environmental conditions and social expectations."

[15] Osonde A. Osoba and William Welser IV, "The Risks of Artificial Intelligence to Security and the Future of Work," RAND DOI: 10.7249/PE237, document number: PE-237-RC, 2017, accessed at https://www.rand.org/pubs/perspectives/PE237.html

16 "Home and Away: DHS and the Threats to America,"
    Remarks delivered by Secretary Kelly at George Washington
    University Center for Cyber and Homeland Security, 18 April
    2017, accessed at https://www.dhs.gov/news/2017/04/18/
    home-and-away-dhs-and-threats-america

17 Andy Greenberg, "Over 80 Percent of Dark-Web Visits
    Relate to Pedophilia, Study Finds," *Wired*, 30 December
    2014, accessed at https://www.wired.com/2014/12/80-
    percent-dark-web-visits-relate-pedophilia-study-finds/

18 Although Silk Road was shut down by the FBI, other
    Dark Web marketplaces continue to operate; one 2015
    study estimates the Dark Web economy is between one
    hundred dollars and 180 million dollars annually.

19 Personal and financial information that has been stolen
    via cyberattack ends up for sale on Dark Web markets.
    Tax information, medical files, credit scores and card
    numbers, social security cards, passports, driver's licenses,
    and stolen passwords are all available.

20 Ted Koppel, page 35

21 Hannah Bryce, "The Internet of Things Will Be Even
    More Vulnerable to Cyber Attacks," Chatham House
    Expert Comments, 18 May 2017, accessed at https://
    www.cybersecurityintelligence.com/blog/the-internet-
    of-things-will-be-even-more-vulnerable-to-cyber-at-
    tacks-2456.html

22 Jeffrey Lewis, "Our Nuclear Future: Looking into the
    Abyss; Is Nuclear War Still a Possibility?" *The American
    Scholar,* Summer 2017, page 18, accessed at https://
    theamericanscholar.org/our-nuclear-future/#

23 United Nations, Department of Economic and Social
    Affairs, page 6: "Eighty-two percent of cities, representing
    approximately eighty-eight percent of the total city popu-

lation, were exposed to high mortality vulnerability from at least one type of natural disaster, and nearly ninety percent of cities, representing more than ninety-three percent of the total city population, were highly vulnerable to economic losses from at least one type of natural disaster."

24 "The Impact of Climate Change on Natural Disasters," NASA Earth Observatory, webpage accessed at https://earthobservatory.nasa.gov/Features/RisingCost/rising_cost5.php

25 Barbara Whitaker, "Ready for Anything (That's Their Job)," *The New York Times*, 9 September 2007, accessed at http://www.nytimes.com/2007/09/09/jobs/09starts.html

26 Lori Uscher-Pines, Anita Chandra, Joie Acosta, and Arthur L. Kellermann, "Why Aren't Americans Listening to Disaster Preparedness Messages?" The Rand Blog, June 29, 2012, accessed at http://www.rand.org/blog/2012/06/why-arent-americans-listening-to-disaster-preparedness.html: "Two-thirds of Americans don't believe they will be on their own after a disaster—they think that first responders will arrive within seventy-two hours. Many low-income individuals—the group usually hardest hit by a disaster—don't have enough money to stock a disaster kit or a place to store it."

27 Ad Council, "Real Stories: Emergency Preparedness," 2018, accessed at http://www.adcouncil.org/Impact/Real-Stories/Emergency-Preparedness

28 Deborah Wilson, "'I'll be OK' attitude behind lack of disaster preparedness, study finds," CBC News, 24 January 2018, accessed at http://www.cbc.ca/1.4502644

29 Nassim Nicholas Taleb, *The Black Swan: The Impact of the Highly Improbable*, Random House, 2007

30 War stories were against OEM meeting rules and the fine was 1 dollar per war story. You could crowdfund by placing a hat in the middle of a conference table during a typical meeting.

31 Including individuals and families (including those with access and functional needs), businesses, faith-based, and community organizations, nonprofit groups, schools and academia, media outlets, and state, local, tribal, territorial, and the federal government.

32 The dictum of Socrates: "The only thing I know is that I do not know."

33 Think of earthquakes.

34 Ted G. Lewis, "Cause-and-Effect or Fooled by Randomness?" *Homeland Security Affairs*, vol. VI, no. 1 (January 2010), www.hsaj.org

35 Nassim Nicholas Taleb, *Fooled by Randomness: The Hidden Role of Chance in Life and in the Markets*. Penguin, 2013

36 Patrick Hlucy, "The man behind Murphy's Law," *Toronto Star*, 11 January 2009, accessed at https://www.thestar.com/news/2009/01/11/the_man_behind_murphys_law.html

37 Per Bak, How nature works: the science of self-organized criticality, Springer-Verlag New York, Inc., 1996

38 Tia Ghose, "Hurricane Harvey Caused 500,000-Year Floods in Some Areas," Live Science, 11 September 2017, accessed at https://www.livescience.com/60378-hurricane-harvey-once-in-500000-year-flood.html: "In 2017, Houston received a one-in-a-thousand-year flooding over a day, while the total rainfall over the five-day period reached the one-in-five-hundred-thousand-year mark in some small pockets from Hurricane Harvey."

39 Charles Perrow, "Normal Accident at Three Mile Island," *Society*, July/August 1981, accessed at http://www.penelopeironstone.com/Perrow.pdf: "A minor malfunction in a cooling circuit in reactor number two (TMI-2) caused the temperature in the primary coolant to rise. This in turn caused TMI-2 to shut down automatically. At this point a relief valve failed to close, but the control panel did not reveal it, so much of the primary coolant drained away and TMI-2 began to melt down. The operators were unable to diagnose or respond properly to the automatic shutdown of the reactor. Postincident investigations pinpoint deficient control room instrumentation and inadequate emergency response training as the root causes of the accident."

40 Ibid, page 23, "The cause of the accident is to be found in the complexity of the system."

41 Jeffrey A. Lewis, page 18

42 Nassim Nicholas Taleb, *The Black Swan: The Impact of the Highly Improbable*

43 Lewis Lapham, "Petrified Forest, Fear is America's top-selling consumer product," *Lapham's Quarterly*, 10 July 2017, accessed at https://www.laphamsquarterly.org/fear/petrified-forest

44 Otto Hahn, Lise Meitner, and Fritz Strassman

45 Led by Major General Leslie Groves of the US Army Corps of Engineers and nuclear physicist Robert Oppenheimer.

46 Jeffrey Lewis, page 18

47 Jeff Nuttall, *Bomb Culture*, Paladin, 1972. The profound effect of the bomb on teenagers was examined by Nuttall in his 1968 survey of postwar youth culture.

48 Jon Savage, "Pop in the age of the atomic bomb," *The Guardian*, 31 October 2010, accessed at https://www.

theguardian.com/music/2010/oct/31/pop-music-atomic-bomb-jon-savage

49 *Duck and Cover* (the film), Wikipedia, accessed at https://en.wikipedia.org/wiki/Duck_and_Cover_(film)

50 Wikipedia, Emergency Broadcast System, accessed at https://en.wikipedia.org/wiki/Emergency_Broadcast_System: "From 1963 to 1997, EBS provided 'the President of the United States with an expeditious method of communicating with the American public in the event of war, threat of war, or grave national crisis.' "This is a test of the Emergency Broadcast System. The broadcasters of your area in voluntary cooperation with the FCC and other authorities have developed this system to keep you informed in the event of an emergency. If this had been an actual emergency, you would have been instructed where to tune in your area for news and official information. This is only a test. This concludes this test of the Emergency Broadcast System."

51 Susan Roy, *Bomboozled: How the U.S. Government Misled Itself and Its People into Believing They Could Survive a Nuclear Attack*, Pointed Leaf, 2011: "Many groups—including the Catholic Worker Movement—actively protested the program, claiming it 'misled the public that it could survive a nuclear war.' Although rarely done in practice, opposition to evacuation drills increased during the 1950s until they were finally discontinued in 1961. As the Cold War dragged on into the 1970s and '80s, civil defense continued to be plagued by fits and starts."

52 Jeffrey Lewis, webpage

53 Drew Atkins, "Washington is 'shockingly unprepared for a nuclear attack," Features.crosscut.com, 19 July 2017: For the past half century, though, public perception

that there is no hope for survival has been an impediment to planning for it. In some places it has even been banned. A law passed in Washington state in 1984 says that state disaster plans "may not include preparation for emergency evacuation or relocation of residents in anticipation of nuclear attack." Not surprisingly, the state of Washington was recently accused of being "shockingly unprepared for a nuclear attack." Accessed at http://features.crosscut.com/washington-is-shockingly-unprepared-for-a-nuclear-attack

54 Steven T. Ganyard, "All Disasters Are Local," *The New York Times*, 17 May 2009, accessed at http://www.nytimes.com/2009/05/18/opinion/18ganyard.html

55 In addition to tribal governments, as of 2003 there were 562 federally recognized Indian nations in the US.

56 Stephan Kozub, "Take a deep breath — no really, it will calm your brain," *The Verge*, 30 March 2017, accessed at https://www.theverge.com/2017/3/30/15109762/deep-breath-study-breathing-affects-brain-neurons-emotional-state

57 Disaster professionals call this essential first question "evacuation versus shelter in place."

58 Right now the voice in your head is complaining loudly about how boring all of this is. Recognize that this is your mind reacting to so much thinking. It has never had to work nearly this hard outside the parallel universe.

59 Irene Langridge, *William Blake: A Study of His Life and Art Work*. Hardpress Publishing, 2012: "O glory! O delight! I have entirely reduced that spectrous Fiend to his station, whose annoyance has been the ruin of my labours."

[60] In the parallel universe, the most important tool in your toolbox is your mind.

[61] When the truth is just the opposite, "you can handle it. You can figure it out. Life as you know it is not over. Everything is going to be all right."

[62] Leonard J. Marcus, Ph.D.; Isaac Ashkenazi, M.D.; Barry Dorn, M.D.; Joseph Henderson, M.A.; and Eric J. McNulty, *Meta-Leadership: A Primer*, 2010

[63] It turns out that one of your employees who was killed was a dog lover who lived alone. A colleague went to Alex's house to check on him and his dogs and found the front door wide open and the three Labradors gone. News of this triggers a fresh wave of panic among your staff.

[64] The day Hurricane Katrina made landfall.

[65] This excludes first responder services such as firefighting, law enforcement, search and rescue, and emergency medical services who are in the field 24/7.

[66] We call this extreme multitasking.

[67] Sun Tzu, *Art of War,* William Collins, 2018

[68] I call it the "why does it feel like we are all doing this for the first time?" feeling.

[69] From their first day, police officers are trained in how to react when fired upon. Yet when that first shot is fired, some will not recognize it. In the first moments, their mind will play tricks: "It's kids playing" or "This is a movie shoot."

[70] Kathryn Schulz, "The Really Big One," *The New Yorker*, 20 July 2015, accessed at http://www.newyorker.com/magazine/2015/07/20/the-really-big-one

[71] Ibid, page 11

[72] Robert Ellsworth Wise, Jr.

73 This excludes first responder services such as firefighting, law enforcement, search and rescue, and emergency medical services who are in the field 24/7.

74 Carroll Doherty, Jocelyn Kiley and Olivia O'Hea, "Government Gets Lower Ratings for Handling Health Care, Environment, Disaster Response, Low Trust in Federal Government Among Members of Both Parties," Pew Research Center, 14 December 2017, http://www. people-press.org/2017/12/14/government-gets-lower-ratings-for-handling-health-care-environment-disaster-response/2/: "Large majorities say the government should play a major role in responding to natural disasters (89 percent). Somewhat smaller majorities (64 percent) say the government is doing at least a somewhat good job in this area."

75 Daniel Bliss, *Economic Development and Governance in Small Town America: Paths to Growth.* Routledge, 2018

76 Article I, Section 8 of the US Constitution specifies the "expressed" or "enumerated" powers of Congress. These specific powers form the basis of the American system of federalism, the division and sharing of powers between the central government and the state governments.

77 It turns out that Article I, Section 8 of the Constitution says nothing about Congress having the power to subsidize or pay for disaster relief.

78 Including counties, municipalities, towns or townships, or villages.

79 And Dillon's Rule.

80 Except New Mexico, Utah, and West Virginia.

81 Voice of America, "Trump Vows Quick Response to Hurricanes as Forecasters Predict High Number of Storms," *VOA News*, 4 August 2017, accessed at

https://www.voanews.com/a/trump-visits-federal-agency-for-briefing-on-hurricane-season/3972279.html

[82] Yxta Maya Murray, "In Puerto Rico, the 'natural disaster' is the US government," 23 November 2017, *The Hill*, accessed at http://thehill.com/opinion/international/361235-in-puerto-rico-the-natural-disaster-is-the-us-government

[83] The most active Atlantic hurricane season in recorded history, with five major hurricanes making landfall in the US—Dennis, Emily, Katrina, Rita, and Wilma.

[84] Faucett, Richard, "Amid Chaos of Storms, U.S. Shows it Has Improved its Response," Richard Faucett, *The New York Times*, 12 September 2017, accessed at https://www.nytimes.com/2017/09/12/us/irma-harvey-hurricane-response.html

[85] Wikipedia, Hurricane Maria entry, accessed at https://en.wikipedia.org/wiki/Hurricane_Maria

[86] Alan Gomez, "Trump's wildly different responses to hurricanes in Texas, Florida and Puerto Rico," *USA Today*, 12 October 2017, accessed at https://www.usatoday.com/story/news/world/2017/10/12/tweets-visits-threats-how-trumpte-takes-different-approach-hurricane-relief-puerto-rico-texas-florid/757201001/

[87] Disasters and Emergencies, World Health Organization/EHA, Panafrican Emergency Training Centre, Addis Ababa, Updated March 2002: accessed at http://apps.who.int/disasters/repo/7656.pdf: "A disaster is an occurrence disrupting the normal conditions of existence and causing a level of suffering that exceeds the capacity of adjustment of the affected community."

[88] Or commonwealth, as in the case of the Hurricane Maria response.

89　David Smith, "Obama praises Fema's 'change of culture' during tour of Louisiana flood recovery," *The Guardian*, 23 August 2016, accessed at https://www.theguardian.com/us-news/2016/aug/23/barack-obama-fema-louisiana-flood-recovery-culture-katrina

90　David Graham, "We Are All First Responders," *The Atlantic*, 3 September 2015, accessed https://www.theatlantic.com/national/archive/2015/09/we-are-all-first-responders/402146/

91　Robert T. Stafford Disaster Relief and Emergency Assistance Act, PL 100-707, signed into law November 23, 1988; amended the Disaster Relief Act of 1974, PL 93-288. This act constitutes the statutory authority for most federal disaster response activities, especially as they pertain to FEMA and FEMA's programs.

92　Assuming typical support through mutual aid, EMAC, and the like.

93　Ken Thomas, "Trump says states can count on federal cash in emergencies," Associated Press, 4 August 2017, accessed at https://www.fema.gov/media-library/assets/documents/15271

94　Under the Stafford Act, disaster-related damages must top approximately 1.4 dollars per capita, which for several states is less than one million dollars. Even local disasters that are centered in one state and cost as little as a million dollars can qualify for a presidential declaration. Accessed at https://www.fema.gov/pdf/media/factsheets/dad_disaster_declaration.pdf

95　Various Authors, "Hurricane Andrew Remembered," accessed at http://www.andrew20th.com/

96　D. Himberger, D. Sulek, S. Krill, "When there is no Cavalry," *Disaster Resource Guide*, 2007–08, p. 52, accessed at http://wp.cune.org/tarasheets/files/2012/11/LO1HPM_HurricaneAndrewEmergencyPreparedness.pdf

97 Tony Adamski, Beth Kline and Tanya Tyrrell, "FEMA Reorganization and the Response to Hurricane Disaster Relief," *Perspectives in Public Affairs*, vol. 3 (Spring 2006), accessed at https://www.asu.edu/mpa/FEMAReorganization.pdf

98 Tracy Hughes, "The Evolution of Federal Emergency Response Since Hurricane Andrew," Fire Engineering and American Public University, 1 February 2012, accessed at http://www.fireengineering.com/articles/print/volume-165/issue-2/features/the-evolution-of-federal-emergency-response-since-hurricane-andrew.html

99 Robert Pear, "Hurricane Andrew; Breakdown Seen in U.S. Storm Aid," *The New York Times*, 29 August 1992, accessed at http://www.nytimes.com/1992/08/29/us/hurricane-andrew-breakdown-seen-in-us-storm-aid.html

100 Stafford Disaster Relief and Emergency Assistance Act, accessed at https://www.fema.gov/pdf/media/factsheets/dad_disaster_declaration.pdf

101 Joseph F. Zimmerman, "Interstate Relations Trends," The Book of the States, Council of State Governments, 2011 accessed at: http://knowledgecenter.csg.org/kc/system/files/Zimmerman2011.pdf: "An Imperium in Imperio (an empire within an empire) is an apt descriptor of a federal system, as sovereign political powers are divided between a national government and constituent state governments. This power division in the United States automatically produces interstate relations characterized by competition, cooperation, and/or controversies."

102 David Inserra, "FEMA Reform Needed: Congress Must Act," Homeland Security Report, Heritage

Foundation, 4 February 2015, accessed at http://www. heritage.org/research/reports/2015/02/fema-reform-needed-congress-must-act

103 Ted Koppel, *Lights Out: A Cyberattack, a Nation Unprepared, Surviving the Aftermath*, New York: Random House, 2015. Print.

104 Transcript: "Brock Long on Face the Nation," 3 September 2017, accessed at https://www.cbsnews.com/news/ transcript-brock-long-on-face-the-nation-sept-3-2017/

105 Abby Phillip, Ed O'Keefe, Nick Miroff, Damian Paletta, "Lost weekend: How Trump's time at his golf club hurt the response to Maria," The Washington Post, 29 September 2017, accessed at https://www.washingtonpost.com/ amphtml/politics/lost-weekend-how-trumps-time-at-his-golf-club-hurt-the-response-to-maria/2017/09/29/ ce92ed0a-a522-11e7-8c37-e1d99ad6aa22_story.html

106 John Scrivani is a former New York City police officer who rose through the ranks to command the hazardous materials division in the elite Emergency Services Unit during the city's response to 9/11. After leaving NYPD, Scrivani joined the New York City OEM as the Deputy Commissioner for operations. This an excerpt from his farewell email sent to OEM on September 2, 2011.

107 Dan Fastenberg, "Sign of the Economic Times: Squeegee Men Return To New York Streets," AOL, 21 September 2011, accessed at https://www.aol.com/2011/09/21/ sign-of-the-economic-times-nyc-squeegee-men-return-to-streets/

108 Except California, Texas, and New York.

109 As one former lieutenant put it, "Rudy walked around with the org chart of the city in his head."

110 And former director of the Indiana Department of Emergency Management.

111 OEM initiated a large-scale citywide planning effort in anticipation of the Y2K Bug, also called the Year 2000 Bug or Millennium Bug, a problem in the coding of computerized systems that was projected to create havoc in computers and computer networks around the world at the beginning of the year 2000.

112 Both written and unwritten.

113 Before Lhota, Giuliani's enforcer and peacemaker was the suspenders-sporting Randy Mastro.

114 His police commissioner, William Bratton, had applied the broken windows theory of policing, which holds that minor violations create a permissive atmosphere that leads to more serious crimes.

115 Gary M. Klass, "NYC crime rate cut with penalties," BCHeights.com, 3 November 2005: "It's the street tax paid to drunks and panhandlers. It's the squeegee men shaking down the motorist waiting at a light. It's the trash storms, the swirling mass of garbage left by peddlers and panhandlers, and open-air drug bazaars on unclean streets."

116 Michael Tomasky, "The Day Everything Changed," New York Magazine, 28 September 2008, accessed at http://nymag.com/anniversary/40th/50652/index1.html

117 Also known as World Trade Center 1.

118 James Glanz, "A Nation Challenged: The Site; Engineers Have a Culprit in the Strange Collapse of 7 World Trade Center: Diesel Fuel," *The New York Times*, 29 November 2001, accessed at http://www.nytimes.com/2001/11/29/nyregion/nation-challenged-site-engineers-have-culprit-strange-collapse-7-world-trade.html

[119] Like the RMS Queen Mary and the SS Normandie.

[120] As well as nonprofit and private sector organizations.

[121] History.com staff, "Reaction to 9/11," The History Channel, published in 2010, accessed at http://www.history.com/topics/reaction-to-9-11

[122] Resembling Purina Dog Chow.

[123] Robert D. McFadden, "A Record Snow: 26.9 Inches Fall in New York City," *The New York Times*, 13 February 2006, accessed at http://www.nytimes.com/2006/02/13/nyregion/a-record-snow-269-inches-fall-in-new-york-city.html

[124] It matters little what we do two weeks into the disaster. Gold-plated Rolls-Royces won't quell the complaints if you are not ready on day one, the golden hours of the disaster.

[125] The red, white, and blue teams, with each on-call for three weeks and then off for six. Since each could operate an emergency operations center, it wasn't necessary to activate the entire agency for every big job. Even though we all knew we would get pulled in if the job were big enough, not having to be on-call fifty-two weeks a year was a huge psychological benefit. We would not have been able to do what we did without the red, white, and blue construct. It allowed us to get a lot of practice but not so much that we were crushed under its burden.

[126] Khalid al-Hammadi, "The Inside Story of al-Qa'ida," Part 4, Al-Quds al-Arabi, 22 March 2005, accessed at https://www.bookemon.com/asset/reader_1/pagereader_8.5_11.swf: "Al Qaeda's motto is 'centralization of decision and decentralization of execution.'"

[127] General Stanley A. McChrystal, Tantum Collins, David Silverman, and Chris Fussell, *Team of Teams: New Rules*

of Engagement for a Complex World. Portfolio/Penguin, 2015, page 219

[128] Ibid, page 226

[129] Ibid, page 232

[130] Written on the giant whiteboard behind the podium at the Harris County, Texas, emergency operations center during the Harris County Office of Homeland Security's and emergency management's epic response to Hurricane Harvey in August and September 2017.

[131] Or commonwealth, as in the case of the Hurricane Maria response.

[132] David Boaz, "Catastrophe in Big Easy Demonstrates Big Government's Failure," Commentary, Cato Institute, 19 September 2005, accessed at https://www.cato.org/publications/commentary/catastrophe-big-easy-demonstrates-big-governments-failure

[133] Select Bipartisan Committee to Investigate the Preparation for and Response to Hurricane Katrina, "A Failure of Initiative: Final Report of the Select Bipartisan Committee to Investigate the Preparation for and Response to Hurricane Katrina," US Government Printing Office, 2006

[134] When the earth, the sun, and the moon are in a line and their gravitational pull is at its maximum.

[135] New York City Special Initiative for Rebuilding and Resiliency, "Sandy and Its Impacts," 11 June 2013, accessed at http://www.nyc.gov/html/sirr/downloads/pdf/final_report/Ch_1_SandyImpacts_FINAL_singles.pdf

[136] Measuring 13.88 feet at the Battery in Lower Manhattan, surpassing the old record of 10.02 feet measured there during Hurricane Donna in 1960.

137 John Homan, "The City and the Storm," New York Magazine, 4 November 2012, accessed at http://nymag. com/news/features/hurricane-sandy-2012-11/

138 That included NYC restoration centers, FEMA's Individual and Public Assistance Programs, the NYC Rapid Repair Program, and NYC Build It Back.

139 *The Faulkner Reader: Selections from the Works of William Faulkner.* Modern Library, 1971

140 As I write this, wildfires continue to burn in the worst wildfire season in Northern California's history. The 2017 California wildfires consisted of 8,176 fires burning 1,079,569 acres, with forty-two dead, 8,400 structures burned, and one billion dollars in damage.

141 Daryl Osby, Ken Kehmna, Firefighting Resources of Southern California Organized for Potential Emergencies (FIRESCOPE), accessed at https://www.firescope.org/ index.php

142 Amanda Griscom, "Man Behind the Mayor," New York Magazine, 15 October 2001, accessed at http://nymag. com/nymetro/news/sept11/features/5270/

143 Brian Michael Jenkins; Frances Edwards-Winslow, Ph.D., CEM, "Saving City Lifelines: Lessons Learned in the 9-11 Terrorist Attacks," MTI Report 02-06, September 2003, accessed at http://transweb.sjsu.edu/MTIportal/ research/publications/documents/Sept11.book.htm

144 The CDC's Strategic National Stockpile is a supply of pharmaceuticals and medical supplies for use in a public health emergency. The Push Package is its first tranche, a fifty-ton cache that can be delivered to anywhere in the US within twelve hours.

145 One of the three bulk samples showed a one percent level of asbestos; the other two samples were negative. The lab

used phase contrast microscopy (PCM) to analyze the air samples and found that there was a fiber count elevation in the air consistent with asbestos. The fiber count from the air sample was approximately 0.06 fibers per cubic centimeter. This is lower than the permissible exposure limit, or PEL, set by OSHA, which is 0.1 fibers per cubic centimeter. When asbestos reaches a concentration of 0.1 fibers per cubic centimeter, OSHA regulations force employers to remove the asbestos or put people into respirators, whichever is more practical.

[146] White polyethylene coveralls, the so-called "moon suits."

[147] In everyday life, DCAS takes six months to approve a new purchase, but red tape melts in the parallel universe.

[148] Any criticism of the EPA cannot extend to its people. EPA field staff were all-in on this with us. Brett Plough (a pseudonym for Steve Touw) worked as hard as we did, maybe harder, and he cared. In my mind, New York City owes Touw and the other on-scene coordinators, such as Mike Soletsky and Jim DaLoia, a debt of gratitude.

[149] The health and safety professionals who went above and beyond the call of duty, working around the clock, on the mission of the WTC Health and Safety Task Force are too numerous to list here. I will name just a few of them: Bob Adams (DDC), Ray Master (Bovis), Stewart Burkhammer (Bechtel), Dan Hewett, Ken Martinez (CDC/ NIOSH), Gil Gillen, Rich Mendelson, Pat Clark (OSHA), Steve Touw, Michael Soletsky, Andrew Confortini (EPA), Michael Mucci (DSNY), Michael Nardone (FBI), Ralph Pascarella (operating engineer), Maureen Cox (PESH), Sylvia Pryce (COSH), Tom Mignone (USPHS), Phil Taylor, Mary Plaskon (Port Authority), Ronald Spadafora

(FDNY), John Scrivani (NYPD), Dr Jim Miller, Mark Foggin, and Allan Goldberg (DOH).

150 On September 20, 2001, at the city's request, OSHA assumed full responsibility for respirator distribution, fit-testing, and training at the WTC site. On September 13, 2001, another comprehensive respirator fit-testing program was established by the New York State Department of Labor Division of Safety and Health. The following day, OSHA also began conducting fit-testing and distributing respirators.

151 Especially Ralph Pascarella.

152 Despite the fact that the World Trade Center site was, in the words of OSHA administrator John Henshaw, "the most dangerous workplace in the United States," in the end there were few serious injuries and no work-related deaths there during the recovery phase.

153 United States Department of Labor, Occupational Health and Safety Administration, "A Dangerous Worksite: The World Trade Center," accessed at https://www.osha.gov/Publications/WTC/dangerous_worksite.htm

154 Robert Neamy, "From Firescope to NIMS," *Fire Rescue Magazine*, 1 August 2011, accessed at http://www.fire-rescuemagazine.com/articles/print/volume-6/issue-8/command-and-leadership/from-firescope-to-nims.html

155 Andrew Freedman, "Panel Finds Flaws with NWS Guidance on Sandy's Surge," Climate Central, 16 May 2013, accessed at http://www.climatecentral.org/news/nws-failed-to-provide-clear-guidance-of-sandys-surge-threat-panel-finds-159

156 European Universities on Professionalization of Humanitarian Action, http://euhap.eu/

[157] Robert D. McFadden, "Carnage on I-95 After Crash Rips Bus Apart," *The New York Times*, 12 March 2011, accessed at http://www.nytimes.com/2011/03/13/nyregion/13crash.html

[158] Ibid, webpage

[159] A passenger manifest is a document listing the passengers of a ship, an aircraft, or another vehicle, for use by customs or other officials. When the last passenger steps onto a commercial airline flight, for instance, the airline "closes," or issues a passenger manifest with name, seat assignment, and emergency contact information for every "soul" aboard. A bus accident is what is referred to as an open-manifest incident, meaning the transporter does not create a record of its passengers.

[160] Dina Maniotis and the human services team at OEM had led a process to build the New York City Family Assistance Center Plan and engage a powerful group of partner agencies, including Office of Chief Medical Examiner, NYPD, Agency for Children Services, American Red Cross, Salvation Army, Tzu Chi, Department for the Aging, Department of Education, Human Resources Administration, Department of Health and Mental Hygiene, US Department of State, NYC Housing Preservation and Development, Mayor's Office of Veteran's Affairs, Mayor's Office for People with Disabilities, MTA, NYC Bar Association, NYC Small Business Services/ US Small Business Administration, NYS Insurance Department, and Animal Care and Control.

[161] Frank DePaulo, assistant commissioner for operations at the New York City Office of Chief Medical Examiner.

[162] As the hermit guru who lives on the mountaintop said, "What is all this fuss and fret about the rules? The rules

are not for you. Police officers and firefighters put their lives at risk everyday. Yet you fail to act because you fear harsh words and reprimands? Rules are for daily life, not for disasters. It is your job to throw them out. If, after the job, a bureaucrat doesn't ask you, 'Why did you break that rule to help those people?' then you were too meek."

163 Manny Fernandez, Lizette Alvarez and Ron Nixon, "Still Waiting for FEMA in Texas and Florida After Hurricanes," *The New York Times*, 22 October 2017, accessed at https://www.nytimes.com/2017/10/22/us/fema-texas-florida-delays-.html

164 FEMA requires all families who need ONA benefits to apply for a loan from the Small Business Administration. But "business" is a misnomer that confuses people. This alone is a major reason so many families do not get help.

165 FEMA Press Release: "Complete an SBA Loan Application to be Considered for Other Assistance," release date: June 15, 2016, Release Number: NR002, accessed at https://www.fema.gov/news-release/2016/06/15/complete-sba-loan-application-be-considered-other-assistance

166 A perfectly understandable action that thousands of people do.

167 The maximum benefit is 33,000 dollars, with the average grant being just under 8,000 dollars.

168 Tzu Chi USA volunteer, "2017 Central Mexico Earthquake, Tiahuac, Mexico," accessed October 29, 2017 at https://www.facebook.com/TzuChiUSA/

169 Jen Chung, "East Village Explosion Building Manager, Who Was Indicted for Manslaughter, Dies At 31," Gothamist, 3 September 2017, accessed at http://gothamist.com/2017/09/03/east_village_explosion_building_man.php

[170] Marc Santora and Al Baker, "East Village Explosion Ignites Fire, Fells Buildings and Injures at Least 19," *The New York Times*, 26 March 2015, accessed at https://www.nytimes.com/2015/03/27/nyregion/reports-of-explosion-in-east-village.html

[171] ReeRee Rockette, "My Experience in the Superdome during Hurricane Katrina," http://www.rockalily.com/blog/

[172] "A Failure of Initiative, Final Report of the Select Bipartisan Committee to Investigate the Preparation for and Response to Hurricane Katrina," Union Calendar no. 00, 109th Congress, second session, Report 000-000, accessed at http://www.gpoacess.gov/congress/index.html

[173] According to Morales, she was forced to return home and remain there during the hurricane. "I was very scared waiting outside the shelter while they tried to find the key because the wind was picking up and I did not know what I should do." Morales alleged that she still does not know which shelters, if any, will be accessible during an emergency.

[174] Post Staff Report, "Planning Makes Perfect," *New York Post*, 30 August 2011: "Credit goes to the Coastal Storm Plan—or, more to the point, to the people who prepared it so precisely and then executed it so flawlessly" accessed at http://nypost.com/2011/08/30/planning-makes-perfect/

[175] Judge Furman: "Given [its] challenges, and what New York City has had to face in recent years, the city's planning and response have been remarkable in many ways. In particular, the array and detail of its plans for every imaginable kind of emergency is impressive, and the valor and sacrifice of those who have come to the aid of New Yorkers in times of emergency, from first responders

to volunteers, have been nothing short of extraordinary. This lawsuit does not challenge those facts. Far from it: In many respects, this lawsuit has confirmed them." He also said, "The record in this case makes clear that, although the city's emergency preparedness plans fall short of legal requirements in several significant respects, they are still remarkable in many ways. The challenges facing cities in general, and this city in particular, are immense, and New York City has done an admirable job of preparing for a wide range of disasters, both man-made and natural, that could strike at almost any time."

176 *Collected Works of Abraham Lincoln.* Volume 1., Lincoln, Abraham, 1809-1865, University of Michigan Press, accessed at https://quod.lib.umich.edu/l/lincoln/lincoln1/1:498

177 That corner, the intersection of Chambers and West Streets in Lower Manhattan, is three blocks north of the World Trade Center. It was the location of the final security checkpoint in the red zone, the "gateway to Ground Zero."

178 C-SPAN, New York City news conference video. "New York City officials held a news conference after a driver ran down people, killing at least eight and injuring others," 31 October 2017, accessed at https://www.c-span.org/video/?436656-1/new-york-city-news-conference

179 Alan Henry, "How to Prepare Your Community for a Disaster," *The New York Times*, February 15, 2018, accessed at https://www.nytimes.com/2018/02/15/smarter-living/prepare-your-community-for-a-disaster.html

180 Shankar Vedantam, "The Key to Disaster Survival? Friends And Neighbors," All Things Considered, 4 July 2011, accessed at https://www.npr.org/2011/07/04/137526401/the-key-to-disaster-survival-friends-and-

neighbors: A research study in the visited villages in India hit by the giant 2004 tsunami found that villagers who fared best after the disaster weren't those with the most money or the most power. They were people who knew lots of other people—the most socially connected individuals. In other words, if you want to predict who will do well after a disaster, look for faces that keep showing up at all the weddings and funerals. "'Those individuals who had been more involved in local festivals, funerals, and weddings, those were individuals who were tied into the community, they knew who to go to, they knew how to find someone who could help them get aid,' Aldrich said."

[181] Gordon B. Hinckley

[182] General Stanley A. McChrystal, *Team of Teams: New Rules of Engagement for a Complex World.* Portfolio/Penguin, 2015

[183] President John F. Kennedy, "Address at Rice University on the Nation's Space Effort," delivered at Rice Stadium, Houston, Texas, 12 September 1962

[184] North American Aerospace Defense Command is a combined organization of the United States and Canada that provides aerospace warnings and protection for Northern America.

[185] They used national assets; for example, Interagency Modeling and Atmospheric Center (IMAAC), Federal Radiological Monitoring and Assessment Center (FRMAC), Radiological Assistance Program (RAP) teams, and the Radiological Emergency Response Team to acquire key incident data and generate fallout predictions.

[186] Including the Emergency Alert System and wireless emergency alerts, wireless communications including UHF and VHF radio frequency (for example, 400- and 800-megahertz radio) and amateur (ham) radio.

about the author

Kelly McKinney has had a leadership role in every major disaster in New York City for more than fifteen years, from the aftermath of the 9/11 attacks to the present day.

As Deputy Commissioner at the New York City Office of Emergency Management, he led the city's response to Hurricane Sandy. He assembled a Task Force of hundreds, led by the FDNY and the National Guard, to knock on thousands of doors across the coastal areas devastated by the storm. He rebuilt the city's disaster planning program, including the Coastal Storm Plan, about which the *New York Post* said, "Planning makes perfect."

As Chief Disaster Officer for the American Red Cross he rushed to the aid of people affected by train crashes and building collapses, most notably the March 2014 Con Edison gas explosion on 125th Street in Harlem.

He is the founding principal of Emergency Management Americas, a 501c3 nonprofit with a mission to advance the profession and practice of emergency management.

Nationally known for his writing and speaking on the principles and practice of disaster management, his five-minute video, *The Essential Emergency Manager*, has been viewed more than 70,000 times on YouTube.

He is a professional engineer with a BS in mechanical engineering from the University of Kansas and an MPA from Columbia University in the City of New York.

He is a board member of the All-Hazards Consortium and of the Urban Assembly School for Emergency Management in New York City.

Currently, he is the Senior Director of Emergency Management and Enterprise Resilience at NYU Langone Health, a world-class academic medical center based in New York City.